SCIENCE FOR EXERCISE AND SPORT

Science for Exercise and Sport is a handbook written for undergraduate sport studies and sport and exercise students. It introduces students to the basic scientific principles that will underpin their learning during their studies and is aimed primarily at students who have limited knowledge in science.

In this book, Craig Williams and David James relate key scientific concepts to an applied situation in order to help gain an understanding by reflecting on the applied exercise and sport examples.

The text is divided into three sections: the first part covers the three physical states of matter (gas, liquid and solid); the second part explains forces, energy and electricity, covering topics that include pressure, torque and power; and the final section focuses on data analysis, ICT and report writing – important areas for the scientist.

Craig Williams is Programme Leader and Lecturer in Exercise and Sport Sciences at the University of Exeter. **David James** is Field Chair for the Sport and Exercise Sciences field and lecturer in Exercise Physiology at the Cheltenham and Gloucester College of Higher Education.

SCIENCE FOR EXERCISE AND SPORT

Craig A. Williams

David V. B. James

London and New York

First published 2001
by E &FN Spon, an imprint of Routledge
11 New Fetter Lane, London EC4P 4EE

Simultaneously published in the USA and Canada
by Routledge
29 West 35th Street, New York, NY 10001

Routledge is an imprint of the Taylor & Francis Group

© 2001 Craig A. Williams and David V. B. James

Typeset in Times by
Prepress Projects Ltd, Perth, Scotland
Printed and bound in Great Britain by
TJ International, Padstow, Cornwall

British Library Cataloguing in Publication Data
A catalogue record for this book is available
from the British Library

Library of Congress Cataloging in Publication Data
Williams, Craig A., 1965–
Science for exercise and sport/Craig A. Williams and David V. B. James.
p. cm.
Includes bibliographical references and index.
ISBN 0-419-25160-X – ISBN 0-419-25170-7 (pbk.)
1. Sports sciences – Study and teaching (Higher) – Great Britain. 2. Science –
Study and teaching (Higher) – Great Britain. 3. Physiology – Study and
teaching (Higher) – Great Britain. I. James, David V. B., 1971–. II. Title.
GV558.W55 2001
502'. 4'796–dc21   00-042484

CONTENTS

CONTENTS

FIGURES

TABLES

PREFACE

The number of universities in the United Kingdom which offer undergraduate courses in exercise and sport has increased considerably since the 1980s. There are over 60 different institutions around the country that offer a combination of single or joint honours degrees in Exercise and Sport Science, Coaching Science, Exercise and Health Science, Sport Development, Physical Education, Sport Pedagogy, Sport and Management, Sport and Business Administration or Sport and Computing, to name a few. The increase in courses related to exercise and sport reflects not only the interest students have in the subject, but also the expanding career opportunities in the exercise and sport industry.

The idea for this book has grown out of our experiences as lecturers in exercise and sport science at our respective academic institutions. We are aware that during the first year of undergraduate studies the range of knowledge and skills brought to the lecture theatre or laboratory is very varied, particularly in relation to the student's experiences in science. In some schools at GCSE level, for example, the science subjects are taught through the separate disciplines of chemistry, physics and biology. In other schools it is common practice to combine the science disciplines together and students obtain a GCSE in combined sciences. After studying science some students' last exposure to the subject was their GCSE experience. Our concern as lecturers was therefore how best to facilitate student's learning in the sciences when there is an absence of core knowledge. At the time of writing this book there are very few texts written for the first year undergraduate in exercise and sport. Although there are many quality texts which are devoted to single disciplines within exercise and sport (e.g. biomechanics, psychology, physiology, sociology), there is a gap in texts that cover material which would provide a base of science knowledge within the discipline of exercise and sport.

It is not uncommon for a first year undergraduate student on an exercise and sport degree course to take laboratory-based exercise and sport science modules. This would certainly be the case for a student studying Exercise and Sport Science as a single honours subject, but might also be common for students on a joint honours course who can opt for laboratory-based modules. One of the difficulties of teaching a laboratory-based module is the diverse scientific background of the students. This book was written to help students who enrol for laboratory-based scientific modules in exercise and sport who may not possess a post-16 science qualification. A knowledge of science at the GCSE level has been assumed as the starting point, and, as with all good learning and teaching strategies, it is hoped that enthusiasm for the subject will allow the student to overcome initial weaknesses.

The material covered in the text might be considered typical for a first year undergraduate, although this will vary between courses depending on the specific aims and objectives of the course. It is possible that certain sections of the book might be useful for second or third year students, and skills such as report writing will always be in demand at all levels.

The philosophy behind the book centres around the fact that the student will have a willingness, because of their interest in exercise and sport, to increase their understanding of the science underpinning this discipline. We recognise that no single book could replace a post-16 education in sciences. We have tried not to assume core knowledge, and have concentrated on areas of science that students might experience in a exercise and sport course.

We have approached each topic by initially covering the science that underpins it. It is our belief that in order to improve the understanding of exercise and sport, the scientific knowledge base must be increased prior to application. Once these underpinning scientific principles have been covered, they are then related to an exercise or sport situation. In this respect, students can see the role science has to play in an applied setting, and are then encouraged to think of further applications.

It should also be stressed that this textbook is not a 'catch-all' science book. There are many more principles that we could have covered, but instead we hope that the student will be motivated to go on to read other specialist science books. Our goal has been to make the book as 'user friendly' as possible. Throughout the book we have included a series of action and summary points which we hope the student will find helpful. In addition, we provide a bibliography and a further reading list at the end of most chapters. Throughout the text we have used bold type for key words and these have been included in a glossary.

There are a number of common features throughout the book:

- In each chapter scientific facts are presented first, followed by their application to exercise and sport.
- Each chapter covers a separate topic, but related chapters are grouped into three sections to promote cross referencing.
- Key words are highlighted (in bold) throughout the text and definitions are provided in the glossary.
- Action points are included to assist the student in developing their science knowledge and its application.
- Towards the end of each chapter key points are provided for quick reference.
- At the end of most chapters a bibliography and further reading list are provided. Both lists include text books, journals and official web sites.

In Part 1 the three physical states of matter (gases, liquids and solids) are considered in three separate chapters. In Chapter 2, the defining features of gases are discussed, followed by laws which apply to gases. In the application section, reference is made to the atmospheric gas inspired and the issues surrounding the collection and analysis of expired gas. The various methods that can be used to examine pulmonary gas exchange are also examined.

In Chapter 3 the properties of liquids are considered. Although it is possible to consider gases and liquids under the banner of 'fluids', we have chosen to consider each state separately. The liquid laws are considered, followed by the application of these laws to sport and exercise. The concepts of pressure, buoyancy and flow are examined, and the laws of Archimedes, Pascal and Bernoulli are provided. This scientific basis is then applied to water skiing and blood pressure examples.

Chapter 4 focuses on the properties of solids. The internal energy of solids is examined along with the way in which solids are affected by temperature. Additionally, heat transfer, and factors such as thermal conductivity and radiation, are explained. Heat transfer from a solid object and the structure of bone provide relevant exercise and sport examples.

In Part 2 we look at forces, energy and electricity, which are considered in three chapters (5, 6 and 7 respectively). In Chapter 5 the effect of application of force is discussed in relation to Newton's laws of motion. The relationship between force and pressure is examined, followed by a definition of each term. Different types of forces are discussed, and the measurement of both force and pressure is considered. Applications examine the concept of torque, and the relationship between torque about a joint and joint angle, and torque about a joint and joint movement velocity.

In Chapter 6 the concepts of energy, work and power are defined. Rather than concentrate on energy transfer in the body, a subject that is covered in many physiology and nutrition texts, the chapter focuses on the application of energy, work and power principles to sport and exercise. Examples of pole-vaulting, running and cycling are provided.

Chapter 7 focuses on electricity, including considerations of static electricity, Coulomb's law, the flow of electricity and how electricity relates to power. The applications to exercise and sport include examples such as an electrocardiogram, a defibrillator and bioelectrical impedance analysis.

Part 3 concentrates on data analysis, information and communication technology and report writing and these are examined in three chapters (8, 9 and 10 respectively). In Chapter 8 the process of analysing data is covered, in conjunction with related issues such as hypothesis testing, experimental design, reliability and validity. Throughout the chapter, exercise and sport examples are used. To a large extent the chapter examines the concepts of data analysis rather than the derivation of complex formulae. All students of exercise and sport have to embark on an independent study, project or dissertation at some stage during their course, so this chapter will be useful not only for first years, but for other years as well.

Chapter 9 examines the potential benefits of information and communication technology in the study of exercise and sport. In addition to looking at the range of software applications that a student of exercise and sport might use, the chapter considers the personal computer as part of a network. The advantages of the internet are explored in relation to the needs of a student of exercise and sport.

Chapter 10 focuses on report writing. Laboratory reports are commonly used as a method of dissemination of information from experiments. The ability to write succinctly and scientifically is an important skill. All areas of a report are addressed, from the abstract through to the references. Once again, this chapter will be useful not only for first year students unaccustomed to scientific report writing, but also for second and third year students producing a dissertation.

Craig A. Williams
David V. B. James
August 2000

ACKNOWLEDGEMENTS

We would like to acknowledge input from colleagues at the University of Exeter and Cheltenham and Gloucester College of Higher Education, and at the University of Brighton, where we both spent several years. We are also grateful to the students we have taught over the years, many of whom provided the inspiration for this book.

Part 1

PHYSICAL STATES

1

INTRODUCTION

AIMS OF THE CHAPTER

This chapter aims to introduce science as applied to sport and exercise. After reading this chapter you should be able to:

- understand the basis for a scientific approach;
- recognise the Système International for units of measurement;
- understand the importance of accuracy of measurement;
- recognise scientific notation;
- be aware of issues surrounding measurements on human participants.

Introduction

The study of sport and exercise draws upon many disciplines, including biomechanics, physiology, psychology, medicine, nutrition, philosophy and sociology to name a few. In the sport performance area, biomechanists, physiologists, psychologists, medics and nutritionists often work alongside athletes and coaches with the aim of improving performance. In the exercise and health area, these scientists work alongside patients/healthy clients and exercise and health practitioners with the aim of improving health. In the continuum shown in Figure 1.1, the exercise or sport scientist is portrayed as remote from the other partners in the exercise or sport team.

The perception of the sport or exercise scientist role as a remote one is not useful. It is more useful to see each member of the sport or exercise team as a fully integrated partner with the same common goal (see Figure 1.2). For the sport performers, the goal may be an optimal performance. For a patient or healthy client, the goal may be improved health.

Much of the work of the sport or exercise team should have a scientific basis. If this is the case, members of the team will apply strategies that have proved to be useful. It is the responsibility of the sport or exercise scientist to ensure that the approach has a scientific basis. It has been said that 'There is no such thing as applied science only the application of science' (Huxley). This quotation applies to the area of exercise and sport science, in that to investigate an exercise or sport problem scientific principles are often applied to the problem. In this respect, appreciating what a scientific approach may bring to a situation may be of benefit to all members of an exercise or sport team.

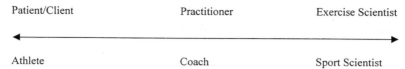

Figure 1.1 A continuum of members of a sport or exercise team

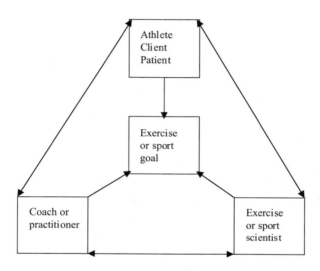

Figure 1.2 An integrated perspective on members of a sport or exercise team

The nature of science

A scientific approach to the study of sport and exercise is often thought of as a useful approach, particularly when the benefits of sport or exercise must be maximised. A involves posing a problem, often in the form of a question, which can then be investigated. Once the question has been investigated, the results may be used to develop a **theory**. The theory may then be applied to similar problems in the future. However, theories evolve as they are challenged. If it is found that the theory does not apply to a certain situation, the theory must be modified to account for that situation. It is the responsibility of scientists to continually challenge theories so that they can be either modified or discarded.

A scientific approach is one that seeks to challenge existing knowledge through the collection of factual information. Prior to the collection of such information, it is necessary to formulate a precise question. The question is normally posed in the form of **hypotheses** that can be tested through the collection of information. Two hypotheses are normally formulated, based on two possible outcomes. One hypothesis is known as the null hypothesis, and relates to a theory remaining unchanged. The other hypothesis is known as the alternative hypothesis, and relates to a need for modification of a theory.

The information that is collected in order to test the hypotheses is normally referred to as **data**. In Chapter 8 the close relationship between data collection, data analysis and hypothesis testing is examined in detail.

Measurement

The collection of data often involves the measurement of some related phenomenon. Measurement might appear an easy task, but on closer inspection it is considered to be an involved process. Data collection depends largely on the quality of measurements made, so the remainder of this chapter will examine measurement issues. Measurement issues will underpin many of the following chapters, hence their inclusion at this early stage of the book.

A student of sport and exercise should have a good understanding of measurement issues. A common problem related to the understanding of measurement is described by Paulos (1988). In an attempt to familiarise students with numbers and just what their quantity means, Paulos asked a student how fast (in miles per hour) human hair grows, to which the student replied, human hair doesn't grow in miles per hour! The answer is actually 10^{-8} miles per hour or 0.00000001 miles per hour. In this example, the student failed to realise that miles per hour was a unit of speed.

To understand measurement units, it is necessary to learn about **Système International (SI) units**. The SI units are an abbreviation of the le Système International d'Unités. These standardised units have been developed to promote international cooperation to provide a universally accepted system of measurement. The system not only allows information to be exchanged from different countries but also offers a standardised form for presentation of measurement details. The seven base units of the SI are shown in Table 1.1. It is important that the fundamental units are learnt because other units are subsequently derived from these seven.

The symbols are a mixture of lowercase and uppercase letters which can be confusing. All symbols are written in lowercase roman letters except when the name of a unit is derived from the name of a person, e.g. W to symbolise power in watts. When the unit has a proper name and is written in full the whole word is in lowercase. For example, the unit of measure for pressure is Pa, named after the scientist Blaise Pascal (1623–1662); when written out in full, it should be written as pascal. A common mistake made by students is the writing of the symbol representing kilogramme (for mass) as Kg; this is incorrect as it should appear kg. Recall from Table 1.1 that the uppercase K is the symbol for temperature, named after Lord Kelvin (1824–1907). This is another example of the rule that all symbols should be lowercase except where the unit is derived from a proper name. An exception to this rule is the litre, where it is now common practice to accept it as L. This exception is partly to avoid confusion between the lowercase l and the numeral 1.

A further convention is never to follow a symbol with a period, unless at the end of a sentence, nor to pluralise symbols. Another common mistake is to mix names and symbols

Table 1.1 The seven base SI units

Physical quantity	Base unit	Symbol
length	metre	m
mass	kilogramme	kg
time	second	s
amount of substance	mole	mol
thermodynamic temperature	kelvin	K
electric current	ampere	A
luminous intensity	candela	cd

together, e.g. newton·metre·s⁻¹. The correct style should be N·m·s⁻¹. Students are also sometimes confused as to when to use the solidus symbol (/). With computers it should be possible to learn how to use the symbol period instead, i.e. · raised above the line when there is a product of two units. The solidus can be used but is not considered to be as precise for scientific work as the dot raised above the text line. If the solidus symbol is used then only one per expression should be included, e.g. kg·m/s². The reason for this is because the division is not associative. When two or more units are formed by multiplication or division in text, a multiplication of two units is indicated by a space between two words and never by a hyphen, e.g. newton-metre would be incorrect, with newton metre being the correct style. To indicate in text a division of several units we would write per, rather than use the solidus, e.g. litres per minute rather than litres/minute. When units are reported, it is preferable to use symbols rather than writing out the full name. Whenever numbers are reported with symbols a space between the two should be left, e.g. 400 W not 400W. As always, there is an exception to the rule: the use of the symbol for degree (°). In this instance, there is no space left between the numerical value and the symbol: 40° not 40 ° or 30°C not 30 °C.

When reporting measurement quantities, a zero should always be placed before a decimal, e.g. 0.01 not .01. Decimals are preferable to fractions, e.g. 0.75 is preferred to ³/₄. If possible long numbers should be separated with the use of a space, e.g. 1,300,000 becomes 1 300 000. This is, however, optional with four digits, e.g. 1000 or 1 000.

Derivatives of units in Table 1.1 are shown in Table 1.2. Students should be aware that there are a number of other non-SI units associated with time. Although the SI unit for time is the second (s), day (d) is sometimes accepted for long periods of seconds. In Table 1.3 are the non-SI units for time commonly used in science journals.

Other derivatives from the original seven base units (Table 1.1) such as energy, power and force will be dealt with in greater detail in Chapters 5 and 6.

One of the reasons why the SI was devised was to help report data that were numerically very large or very small. Therefore, a number of prefixes are used to form multiples and sub-multiples of the base units. These prefixes usually change the quantity by a factor of 10^3 or 10^{-3}, but other smaller or larger increments are used (see Table 1.4). The prefix is written without a space before the base unit and the symbols are as shown in Table 1.4. Therefore, a megagram (Mg) is a kilogramme (kg) multiplied by a thousand.

It is often impractical to write a number with several zeros if you are dealing with very large or small numbers. To overcome this issue, scientific notation or powers of ten can be

Table 1.2 Units derived from the seven base SI units

Derived unit	Name and symbol	Derivation from base units
area	square metre	m^2
volume	cubic metre	m^3*
force	newton (N)	$kg·m·s^{-2}$
pressure	pascal (Pa)	$kg·m^{-1}·s^{-2}$ ($N·m^{-2}$)
work, energy	joule (J)	$kg·m^2·s^{-2}$ ($N·m$)
mass density	kilogramme per cubic metre	$kg·m^3$
frequency	hertz (Hz)	s^{-1}

*Although the cubic metre is the correct derivative for volume, the litre (L), or cubic decimetre (dm^3), has been accepted as the reference for volume.

Table 1.3 Symbols for the non-SI units for time

minute	min
hour	h
day	d
week	wk
month	mo
year	y

Table 1.4 Unit multiples

Prefix	Factor	Equivalent	Symbol
atto	10^{-18}	= 1/1 000 000 000 000 000 000	a
femto	10^{-15}	= 1/1 000 000 000 000 000	f
pico	10^{-12}	= 1/1 000 000 000 000	p
nano	10^{-9}	= 1/1 000 000 000	n
micro	10^{-6}	= 1/1 000 000	μ
milli	10^{-3}	= 1/1 000	m
centi	10^{-2}	= 1/100	c
deci	10^{-1}	= 1/10	d
deka	10^{1}	= 10	da
hekto	10^{2}	= 100	h
kilo	10^{3}	= 1 000	k
mega	10^{6}	= 1 000 000	M
giga	10^{9}	= 1 000 000 000	G
tera	10^{12}	= 1 000 000 000 000	T
peta	10^{15}	= 1 000 000 000 000 000	P
exa	10^{18}	= 1 000 000 000 000 000 000	E

used. It is common practice to shift the decimal point to between the digits in the 'one' and 'ten' positions and then multiply by the appropriate number. For example, if you wanted to write the number of thin filaments in a single muscle fibre, which is 64 000 000 000, on several occasions, it would appear very long winded. By using notation, you can convert this number by placing a decimal point between the first two digits 6 and 4, making 6.4, and then by counting the number of places the decimal point has moved. This conversion is now 6.4×10^{10}. Similarly, 0.00000159 becomes 1.59×10^{-6}. The decimal point has been inserted between the 1 and the 5 to create a number lower than 10, and this is normal practice. Because we are dealing with a small number below 1 in this example, the superscript 6 is preceded by a negative symbol (–).

Now that the correct units of measurement and scientific notation have been discussed, it is important to consider the accuracy of a measure (see Appendix 5). The first point to bear in mind about accuracy is that invariably there is always some error in the measurement. No matter how careful the experimenter is, the measurement is never exact. It is the job of the scientist not only to minimise the error, but also to be aware of the range of error. If a stadiometer measures a person's stature as 1.05 m, it is assumed that the stadiometer can measure between 1.0 and 1.1 m. If a third digit is not relaying any additional information then it would be conventional to record the measure as 1.1. If the stadiometer was poorly calibrated and there was uncertainty in the measurement of ±0.1 m, it would be concluded that the stadiometer cannot be trusted to give an accurate reading. The best guidance is to use the following advice. The final number should not have more digits in it than the most uncertain individual measurement. This is akin to saying that a racing team is only as fast as the slowest runner. If betting how fast a team would complete a race, the speed of the slowest runner would be worked into the equation, rather than ignoring the slowest runner and banking on the fastest runners.

It is common for students to 'blindly' trust all figures that a computerised instrument churns out, the so called 'black-box syndrome'. It is not uncommon for students, therefore, to frantically write down all the figures produced by a computer. For example, a computer program is often used to calculate the maximal oxygen uptake. Invariably a figure produced by the computer reads something like 2.765438 L·min^{-1}, which many students then write down. If the student truly understood the nature of the measurement, they would realise that writing down the figure to that many decimal places would imply the instrument could measure sub-atomic levels. Quite clearly the instrument could not measure with that precision and the correct written form would have been 2.77 L·min^{-1} or even 2.8 L·min^{-1}. Unfortunately in one laboratory demonstration a student had written down −2.34529 L·min^{-1}, demonstrating not only their lack of numeracy but also their belief that the person they had been measuring was photosynthesising rather than respiring!

Making measurements on humans

In the area of exercise and sport, most measurements are made with human participants. Very strict guidelines exist about the treatment of human participants when they are involved in a scientific investigation. When measurements are made on human participants, the research is often referred to as clinical research. Such strict guidelines have arisen due to the potential harm that may be caused to a participant during such investigations. A participant may be harmed in a variety of ways, ranging from the use of procedures which damage health to disclosure of sensitive or confidential information.

Guidelines or codes of practice are published in a number of sources, and are generally based on the **Declaration of Helsinki**. This declaration states:

1   Clinical research should be based on the principles of medical research, and should be based on scientifically established facts from laboratory experiments.
2   Clinical research should be conducted only by scientifically qualified persons.
3   Importance of the objective should be in proportion to the inherent risk.
4   Clinical research should be preceded by careful assessment of inherent risk.
5   Clinical research should be approached particularly cautiously when drugs or experimental procedures may affect the personality of the participant.

Most scientific **journals** in the area of exercise and sport provide guidelines that must be adhered to by researchers who conduct investigations which they intend to publish in such journals. For example, the official journal of the American College of Sports Medicine, *Medicine and Science in Sport and Exercise*, regularly publishes recommendations from the Declaration of Helsinki, and requires all researchers to adhere to those recommendations (e.g. American College of Sports Medicine 1998). Additionally, this journal has published a Policy Statement regarding the use of Human Subjects and **Informed Consent** which must also be adhered to. The policy statement is posted on the journal web page (http://www.wwilkins.com/MSSE/).

When undertaking investigations in exercise and sport, approval is normally required from an institution's **ethics committee** prior to the investigation taking place (see Chapter 10). When measurements are made on human participants, prior **health screening** is essential to establish whether physical activity may cause harm to the participant. Such health screening is normally carried out with a **health questionnaire**, an example of which is included in Appendix 1. In addition to health screening, it is normal practice to obtain **informed consent** from a participant (see Appendix 2). Informed consent simply means that the participant has given consent to take part in an investigation about which they have been fully informed. To be fully informed, the participant must be told about the inherent risks and benefits. Whilst verbal or written consent may be obtained, written consent is generally preferred. For investigations undertaken as part of an undergraduate programme in exercise and sport, ethical approval will normally be sought by the tutor. Nevertheless, it is important that students are aware of the issues when making measurements on human participants.

KEY POINTS

- The common SI units from Table 1.1 used in sport and exercise situations are length (m), mass (kg), time (s), amount of a substance (mol) and thermodynamic temperature (K).
- The process of communicating understanding of the written text is an important feature of science writing. Both SI symbols and numbers used should be correct and accurate.
- An instrument is only as accurate as the divisions on its measuring scale. The number of decimal places recorded implies the confidence in the accuracy of the instrument.
- When dealing with human participants, informed consent should be acquired from the participant before measurements are made.
- When measurements are made on human participants, the risk should be carefully considered.
- When making measurements on human participants, the risk may be lessened by prior health screening, often through a pre-activity health questionnaire.

Bibliography

American College of Sports Medicine (1998) Medicine and Science in Sport and Exercise: Information for Authors. *Medicine and Science in Sport and Exercise* 30(12) (appendix I): i–v.

Paulos, J.A. (1988) *Innumeracy. Mathematical Illiteracy and its Consequences*. London: Penguin Books.

Further reading

Morrison, P., Morrison, P., The Office of Eames, C. and Eames, R. (1982) *Powers of Ten. About the Relative Size of Things in the Universe*. New York: Scientific American Library.

Sale, D.G. (1991) Testing strength and power. In *Physiological Testing of the High Performance Athlete*, J.D. MacDougall, H.A. Wenger and H.J. Green (eds), pp. 68–71. Champaign, IL: Human Kinetics.

Young, D.S. (1987) Implementation of SI units for clinical laboratory data. *Annals of Internal Medicine* 106: 114–129.

2

GASES

This chapter aims to provide an understanding of the scientific principles of gases which are relevant to the theory which underpins sport and exercise. After reading this chapter you should be able to:

- state the properties of gases;
- understand how gases are affected by temperature and pressure;
- understand the concept of the ideal gas laws;
- understand the concept of partial pressure;
- understand the relationship between the diffusion of gases and pressure gradients;
- understand the key concepts in relation to flow of gases in the human body;
- appreciate the terms ATPS, BTPS and STPD;
- apply the scientific principles of gases to laboratory experimentation.

Introduction

The nature of gases is such that they will not occupy a fixed shape or a fixed volume unless contained. Gases will always fill a container completely. Hence, a container of gas may not be described as half full or three-quarters full; a gas container is always full. This nature of a gas contrasts with that of a liquid, which does take the shape of a container, and therefore a container of liquid could be described as being half full (see Chapter 3). In the case of a solid, both shape and volume are fixed at a constant temperature and pressure (see Chapter 4). In the next two chapters, the relationship between volume and pressure in liquids and solids is discussed in detail. The volume of liquids and solids does not alter with a change in pressure, in contrast to the volume of a gas, which does depend on the applied pressure.

An understanding of the properties of gases is important for a student of sport and exercise, since gas exchange underpins the generation of energy in the human body through aerobic metabolism. Energy generation is ultimately dependent upon respiration, which is the term for oxygen utilisation and carbon dioxide production. Respiration is constantly taking place within the human body, and the rate of respiration is markedly increased during exercise to support the increased rate of energy turnover.

The behaviour of liquids is relatively difficult to explain and predict, whereas that of gas is easier to explain. Gases possess a number of properties, such as being easily expanded, and this can be explained by the weak attraction between their molecules. Therefore, there are no strong bonds between adjacent molecules, hence the ability of the gas to easily expand.

Gas molecules do not strongly repel each other. In fact, there is very little regular spacing between gas molecules, unlike that seen for solids. Gas molecules are in constant motion and it is for this reason that a gas will completely fill any container in which it is placed. Gases also differ from liquids and solids in their **density** (mass per unit volume), such that the density of a gas is approximately 1/1000th that of a liquid or solid, a fact which highlights the capability of gas molecules to move further apart. The great distances which gas molecules travel and their speed are impossible to calculate on an individual molecule basis, hence an average value of velocity according to the kinetic theory of gases has been formulated. The kinetic theory of gases relates the pressure to the resulting forces of collisions between molecules where any factors which increase the number of collisions will increase the pressure of the gas.

The temperature of a gas can be used as a measure of the energy of motion and, although the gas movement is random, the behaviour of the gas can be observed. For example, a gas which is much hotter than a neighbouring gas will leak faster than the colder gas when both are leaking through similar sized holes. Hence the hotter the gas, the faster the molecules will move.

A gas whose molecules possess high speeds and move randomly can be described as an **ideal gas**. To understand the concept of an ideal gas, only four variables need to be understood (note that SI units are not always used):

1   Pressure symbolised as P; units of measurement, atmosphere (atm), torr (T), pascal (Pa).
2   Volume symbolised as V; units of measurement, litre (L), centimetre cubed (cm^3), metre cubed (m^3).
3   Temperature symbolised as T; unit of measurement, kelvin (K).
4   **Moles** symbolised as n; unit of measurement, moles (mol).

The four variables can be represented in the following equation, which is known as the ideal gas law:

$$PV = nRT,$$
(2.1)

where P is absolute pressure, V is volume, n is number of moles of the gas (equivalent to the atomic mass unit), R is the gas constant (equivalent to 8.32 J·mol^{-1}·K^{-1}), and T is absolute temperature.

From the ideal gas law two important factors need to be acknowledged. First, notice the explicit use of absolute pressure. This refers to the fact that atmospheric pressure must be added to the pressure reading measured on a pressure gauge. Therefore, the pressure gauge reading plus atmospheric pressure (760 mmHg) equals the absolute pressure. This is also similar for the absolute temperature in kelvins. Hence, absolute zero is equal to zero kelvins (0 K). Note that zero degrees Celsius (0°C) is 273.15 K. Therefore, in all gas calculations the temperature must be in kelvins not Celsius. Note that to make the equation $PV = nRT$ work, SI units should be used for P and V, i.e. Pa for pressure, m^3 for V, unless everything is changed for use with other units.

Fortunately there is an abundance of historical evidence from scientific experimentation which has firmly established the three most common laws related to gas, known as the gas laws. These are:

- Boyle's law (for use with constant temperature or isothermal conditions);
- Charles's law (for use with constant pressure or isobaric conditions);
- the pressure law (for use with constant volume or isovolumetric conditions).

Boyle's law

Boyle's law states that 'the pressure of a fixed mass of gas is inversely proportional to its volume, if the temperature is constant':

$$P \alpha 1/V. \tag{2.2}$$

When applied to gas molecules this law means that as the volume decreases, the gas molecules are occupying a smaller space, and there are a greater number of collisions, hence the pressure will increase. Or conversely, as the volume of gas increases there is a larger space available and so fewer collisions occurring, hence the pressure will decrease. This scenario assumes that the temperature and mass of the gas are constant.

From the ideal gas law equation in Equation 2.1, if the molar mass (n) and temperature (T) are constant, the following relationship is revealed:

$$P_1V_1 = P_2V_2, \tag{2.3}$$

where P_1V_1 are the initial pressure and volume, and P_2V_2 are the final pressure and volume.

Equation 2.3 can be used to calculate the pressure and volume of a gas and what the consequences would be should either of these two factors change. For example, if a gas has a volume of 10 mL and absolute pressure of 760 mmHg and the volume is suddenly increased to 30 mL, what is the resulting absolute pressure? The calculation is:

$$P_1V_1 = P_2V_2$$
$$10 \times 760 = P_2 \times 30$$
$$7600/30 = P_2$$
$$P_2 = 253.3 \text{ mmHg}.$$

In the above example it can be seen that for a factor of three increase in volume, there is a proportional factor of three decrease in the pressure. Manipulating the numbers in the opposite direction would show that for a factor of three decrease in volume, there would be a factor of three increase in pressure.

The law holds true only for gas densities that are low. Gases such as hydrogen (H_2), oxygen (O_2), nitrogen (N_2) and helium (He), also known as the permanent gases, will follow this law at normal pressure but will deviate at high pressure when their density is high.

In the sport of sub-aqua, a scuba diver's exhaled bubbles will expand as they rise to the surface of the water. This is because the pressure exerted by the weight of water decreases with a decrease in depth, so the volume increases as the air bubbles rise. The real life scenario of the scuba diver is critical: divers are told not to ascend from a depth of approximately 10 m (33 feet) without exhaling because the air in the diver's lungs will be expanding to about double its volume. Failure to ascend slowly can lead to debilitating conditions such as the bends due to the expansion of nitrogen gas bubbles within the blood.

Charles's law

Charles's law applies to volume and temperature, whilst pressure is held constant, and was formulated by Charles in 1787. It states that 'a volume at a fixed mass and constant pressure is directly proportional to temperature':

$$V \alpha T.$$
(2.4)

Experimentally, this relationship means that as the temperature increases so too will the number of collisions of the gas molecules, hence the volume will increase. For example, a rugby ball which is inflated indoors and is then left outdoors in the warm sunshine will increase in volume as the temperature of the gas inside the bladder of the rugby ball swells. This is a direct result of the sun's heat.

For calculating changes in volume and temperature, the relationship for Charles's law is:

$$V_1/T_1 = V_2/T_2.$$
(2.5)

Therefore, to determine the change in temperature if a gas had an initial volume of 25 mL and absolute temperature of 298 K and the final volume was 50 mL, the following calculation would be performed:

$$V_1/T_1 = V_2/T_2$$
$$25/298 = 50/T_2$$
$$0.0838 = 50/T_2$$
$$50/0.0838 = T_2 \quad .$$
$$T_2 = 596 \text{ K}.$$

Hence from the above calculation it can be shown that a doubling of the volume will result in a doubling of the temperature.

The pressure law

This law applies to temperature and pressure of a gas if the volume and amount of that gas remain constant. It can be shown that doubling the temperature will double the pressure:

$$P_1/T_1 = P_2/T_2$$
(2.6)
$$760/300 = P_2/600$$
$$2.5 = P_2/600$$
$$2.5 \times 600 = P_2$$
$$P_2 = 1500 \text{ mmHg}.$$

An application of this law would be indicated by the fact that gas bottles, which are often used in laboratories, must not be left in hot environments. The increase in temperature of the gas inside the gas bottles could become great enough for the gas bottle to explode!

The movement of a gas from air into a solution

Air flow is very similar to blood flow as discussed in Chapter 3. The gas from the air must be transported through the respiratory tracts to the lungs, where a small fraction of the gas will move into a solution, i.e. blood. The movement of gas molecules is directly proportional to three factors:

1   The pressure gradient of the individual gas.
2   The solubility of the gas in the given liquid.
3   The temperature.

Air, similar to blood, flows from an area of high pressure to areas of low pressure. Therefore, for air to flow from the atmosphere to the lungs means that there must be higher pressure in the atmosphere than in the lungs. It should be pointed out that the movement due to the pressure gradient also applies to the constituent gases such as oxygen (O_2) or carbon dioxide (CO_2). As air consists primarily of O_2 CO_2, nitrogen (N_2) and water vapour, each of these will exert their own pressure. A law, known as Dalton's law, states that 'the total pressure of a mixture of gases is a result of the pressures of the individual gases'. Hence, air at sea level has a total gas pressure of 760 mmHg, which is the sum of the individual gas pressures:

Partial pressure = % concentration/100   × total pressure of gas mixture   (2.7)

Oxygen:

159 mmHg   = 20.93/100   × 760

Carbon dioxide:

0.23 mmHg   = 0.03/100   × 760

Nitrogen:

600.7 mmHg   = 79.04/100   × 760.

Hence, the **partial pressures** (denoted as P or p) for air at sea level are pO_2 =159 mmHg, pCO_2 = 0.2 mmHg and pN_2 = 601 mmHg. There will be slight variations of the actual partial pressures due to the effect of the water vapour. However, the movement of O_2 and CO_2 from a gaseous state to blood will be dependent on the partial pressure gradients existing between the two mediums. The simple diffusion of O_2 and CO_2 in and out of the lungs is a reflection of the pressure gradients: the larger the gradient the greater the gas diffusion.

The solubility of a gas is the ease with which the gas will dissolve in a solution. Carbon dioxide is about 20 times more soluble in water than O_2. This means that CO_2 molecules more readily move into a solution at very low partial pressures. Oxygen, on the other hand, is not very soluble in watery solutions, hence at high partial pressures only a few molecules may dissolve. A clear example of the lack of solubility of O_2 is the fact that in the blood the majority of O_2 is transported in a form bound to haemoglobin rather than dissolved in the watery solution (the plasma) of the blood.

Scuba diving is a good example of how the gas laws can apply when the human body is exposed to increasing pressure as a diver descends deeper and deeper. The exposure of the diver's body to increased pressure is known as **hyperbaria**. The pressure on the body will increase by 760 mmHg (1 atmosphere) every 10 metres in sea water (there is a slight difference with fresh water because of the difference in water density, such that 1 atmosphere is equal to every 10.4 m). Applying Boyle's law to the diver means that the increasing pressure will decrease the gas volume in the body. At the same time, the increasing pressure will increase the total gas pressures in the lung in proportion to the fraction in the air (Dalton's law). Hence, although the lung volume will be decreasing the partial pressure of the gas in the alveoli (part of the lungs) will be increasing during a descent. When ascending a diver must be careful because as the pressure decreases the alveolar gas partial pressure can suddenly decrease and cause a diver to lose consciousness.

The opposite of hyperbaria, discussed above, is **hypobaria** (decreased barometric pressure). This situation is one which is regularly faced by mountaineers when climbing at altitude. A working knowledge of Dalton's law will assist in understanding the decreasing pressure found with increasing altitude and its effect on the partial pressure of gases. At an altitude of 4000 m, equivalent to 13 123 ft, the barometric pressure is 460 mmHg and the partial pressures of O_2, CO_2 and N_2 in inspired air are:

$$p_IO_2 = 460 \times 0.2093 \quad = 96.3$$

$$p_ICO_2 = 460 \times 0.0003 \quad = 0.14$$

$$p_IN_2 = 460 \times 0.7904 \quad = 363.6$$

Total pressure of mixed gas = 460 mmHg.

Avogadro's law

This law states that if different gases have the same volume, temperature and pressure then they also contain the same number of molecules. If this number of molecules is 6.02×10^{23} (which is Avogadro's number) then the gases have one mole of the gas molecules. The result of this law implies that the volume of a mole of gas is independent of the type of gas. When two samples of gas have the same volume, pressure and temperature, they contain the same number of moles and therefore the same number of molecules.

ACTION POINTS

1   Discuss the effect of temperature on a squash ball during a squash match.
2   If the pressure experienced by a scuba diver at 10 m (33 ft) is approximately 2 atmospheres (1520 mmHg) and one mole of air is 0.79 moles of N_2 and 0.21 of O_2, what are the partial pressures of the gases?

APPLICATION OF SCIENCE TO EXERCISE AND SPORT

1 Correcting gas volumes

The rate of ventilation and the volume of a particular gas uptake or output by the body are important measurements in the study of sport and exercise, as often the volume of gas contained in the lungs needs to be examined. The volume of gas exhaled per minute is known as the expired minute ventilation ($\dot{V}_E$). The volume of oxygen taken up by the body each minute is known as oxygen uptake ($\dot{V}O_2$). The volume of carbon dioxide expelled by the body each minute is known as the carbon dioxide output ($\dot{V}CO_2$). The ratio of $\dot{V}CO_2$ to $\dot{V}O_2$ is known as the respiratory exchange ratio (see Appendix 3). Whilst the total volume of gas in the lungs can never be completely exhaled, the total volume which can be forced from the lungs following an inspiration is often measured, and is termed vital capacity (VC).

An understanding of the gas laws is important when expressing volumes of gas in relation to physiology from either a sports or health perspective. Gas within the lungs is at the temperature of the body (usually 273 + 37 K), is under the same pressure that the body is under (i.e. atmospheric or ambient pressure; usually about 760 mmHg at sea level) and is fully saturated with water vapour ($pH_2O = 47$ mmHg). This set of conditions is known as **BTPS**, which stands for **b**ody **t**emperature, ambient **p**ressure and **s**aturated with water vapour. Once gas is expired, it quickly changes to ambient temperature. Usually the ambient temperature is lower than body temperature and therefore the gas cools. The result of the gas cooling is a decrease in volume and a condensation of water vapour. Under these conditions the gas is said to be at **ATPS**, which stands for **a**mbient **t**emperature, ambient **p**ressure and **s**aturated with water vapour. In order to compare gas volumes, a volume may be reported in the **STPD** form, which stands for **s**tandard **t**emperature (273 K or 0°C), standard **p**ressure (760 mmHg) and completely **d**ry of any water vapour ($pH_2O = 0$ mmHg). This standardised form allows comparison of gas volumes under diverse conditions. Since the barometric pressure, temperature and humidity will vary enormously with time, and throughout the world, a standard form is essential for comparisons. However, when gas volumes are used to express either minute ventilation or vital capacity, they are normally reported in a BTPS form. The reason for this is that it is the absolute volume, rather than a standardised volume, that is important when examining ventilation and lung function.

As already mentioned, gas volumes are also used to express the quantity of oxygen uptake ($\dot{V}O_2$) and carbon dioxide output ($\dot{V}CO_2$) by the body each minute. These quantities are always expressed in a standardised (i.e. STPD) form to aid comparisons between laboratories around the world.

Using the equations from the ideal gas laws, it is possible to standardise a volume of gas collected under ATPS conditions (Equation 2.8). For example, resting oxygen uptake under ATPS conditions can range from 250 to 400 mL·min^{-1}. If a gas volume of 250 mL·min^{-1} was collected and it is assumed that the barometric pressure that was exerted on the gas is 770 mmHg, and the temperature was 25°C, then using the ideal gas laws the standardised or STPD volume would be calculated as follows:

$$P_1V_1/T_1 = P_2V_2/T_2, \tag{2.8}$$

where P is pressure, V is volume, T is temperature, and the subscripts 1 and 2 represent ATPS and STPD values respectively.

Rearranging Equation 2.8 gives

$$P_1 V_1 T_2 / T_1 P_2 = V_2$$

$$[(770 - 24) \times 250 \times (273 + 0)]/[(273 + 25) \times (760 - 0)] = V_2$$

$$V_2 = 224.8 \text{ mL·min}^{-1}.$$

Therefore, the new volume under STPD conditions is 224.8 or, rounded up, 225 mL·min^{-1}. As the body temperature is greater than the ambient temperature the correction to STPD has lowered the volume.

2 Measuring gas volumes during ventilation

One method of measuring gas involves collection of a relatively large quantity of expired gas in a special type of bag (known as a Douglas bag) or meteorological balloon for subsequent measurement. Douglas bags or meteorological balloons are used to collect several expired breaths, and can often hold up to 150 L of gas. Once a collection has been made, the volume measurement can take place at the experimenter's leisure. Normally, a series of collections would be made during an experiment, and the measurement of the gas would take place at the end. Obviously, the material used to manufacture the bags must not be permeable to the gas contained within, so that the time period between collection and analysis does not become critical. The analysis of the gas at the end of the experiment includes the collection of all the required information needed to derive the **pulmonary** gas exchange variables. Such information includes the concentration of O_2 and CO_2 in the expired gas, the temperature of the gas, and the volume of the gas. In addition, the ambient temperature and pressure must be determined. The concentration of O_2 and CO_2 in the expired gas is determined with O_2 and CO_2 gas analysers from a small sample from the bag. The temperature and volume of the gas are determined as the gas passes into a spirometer or volume meter. The measurement of gas temperature at this point is important, since the temperature of the gas is needed in order to convert the volume of gas from ATPS to BTPS. The method of correction into a BTPS form has been discussed earlier. The collection of this data allows determination of a range of respiratory variables, including expired minute ventilation ($\dot{V}_E$), O_2 uptake ($\dot{V}O_2$) and CO_2 output ($\dot{V}CO_2$) for the time over which the gas was collected.

Various devices are available to measure volume of expired gas, including a range of spirometers and gas meters. A spirometer is simply a container of fixed volume, which moves relative to a stable surrounding area, depending upon the volume of gas added to or removed from the container. The output from such a device is interpreted by knowing the output from a fixed input volume. More sophisticated spirometers have an output in litres, so no correction has to be performed. An example of a spirometer used extensively for determination of expired gas volumes is the Tissot spirometer. Gas meters are often referred to as dry gas meters, due to the use to which they are usually put. However, with expired gas, the term 'dry' is misleading, since the volume of gas is saturated.

What has been described above is known as an off-line gas collection and analysis system. It is, however, now possible to determine gas volumes and other respiratory variables on-line. With an on-line system, the mouth-piece is usually extended in the direction away from the subject's mouth to accommodate the device for measuring flow rates of gas (i.e. volume per unit time). Various devices have been put to this use, including turbines, hot-wires, pneumotachographs, and ultrasonic flow meters. All such devices can measure flow rates of expired gas from the mouth continuously and can rapidly detect changes in gas flow.

A turbine is like a small propeller that rotates at various speeds depending upon the flow of gas passing it. Turbines have developed greatly over the past few years in terms of reductions in mass, and now provide a practical method of determining gas flow rates. In the past, turbines were relatively heavy, and consequently had a large amount of inertia. This inertia led to inaccuracies in results, particularly when the turbine was used to measure inspired and expired flow. The previously heavy turbines also resulted in resistance to gas flow, which was undesirable.

Hot-wires utilise the cooling effect of gas flowing past an object. If a wire is kept at a constant temperature, regardless of the gas flow rate past it, the energy put into the wire to keep it at that temperature can be constantly monitored. This energy is proportional to the flow rate of gas past the wire.

Pneumotachographs are pressure-measuring devices, which utilise the potential of gas to exert a pressure over an object. The pneumotachograph looks like a grid of thin wire mesh, and presents a fixed surface area perpendicular to the gas flow direction. Since the surface area is fixed, the pressure exerted against the pneumotachograph is proportional to the velocity of gas flow. It is inherent with such a device that the method of detection of gas flow presents a resistance to flow. At high flow rates, this may be very significant.

Ultrasonic flow meters are based on the distortion of an ultrasonic beam. The beam is presented perpendicular to the direction of flow, and the gas is capable of distorting the beam according to the velocity of flow. The magnitude of distortion is related to the velocity of flow. The advantage of such a device is the lack of resistance to flow. It has been suggested, however, that such devices are better for measuring higher flow rates.

3 Measuring gas concentrations

A large proportion of the gas inspired from the surrounding air (inspirate), and the gas expired from the lungs (expirate), consists of three gases, namely, oxygen (O_2), carbon dioxide (CO_2) and nitrogen (N_2). Whilst the three primary gases do not change in the air that is inspired and expired, the relative proportion of each gas does. Inspirate normally consists of 20.94% O_2, 0.03% CO_2 and 79.03% N_2. The relative proportion of each gas in the expirate depends upon a number of factors, especially whether the individual is at rest or not. During exercise, for example, the faster rate of oxygen utilisation by the body for energy production results in lower O_2 values in expirate compared with values at rest. Conversely, the faster rate of CO_2 production by the body due to energy production, amongst other things, results in higher CO_2 values in expirate compared with values at rest. Nitrogen is an inert gas, and is not normally utilised or produced during exercise. In the study of sport and exercise, when the rate of O_2 utilisation and CO_2 production is increased due to rapid energy utilisation, it is necessary to determine changes in the relative proportions of expirate gas constituents.

Three types of analyser are commonly used for determination of O_2 concentration. Mass spectrometers are used to determine O_2 concentration, often in conjunction with breath-by-breath on-line systems. The mass spectrometer is based on the principle of determination of the relative quantity of molecules with specific atomic masses. The direct nature of this determination makes such a system rapidly responsive. The other two types of analyser, more commonly employed in mixing chamber based on-line systems and off-line systems, are para-magnetic and fuel cell analysers. Para-magnetic analysers make use of the magnetic properties of O_2 in determination of the relative partial pressure in the gas sample. Such analysers are the least rapidly responding. Fuel cell analysers are commonly used in all types of gas analysis systems. The relative amount of O_2 in the gas determines the rate of combustion, and consequently the output from the analyser reflects the relative partial pressure of O_2 in the gas sample.

The concentration of CO_2 in expired gas is usually determined via the infrared absorbance method or mass spectrometry. The latter is a direct determination of the relative number of molecules with specific atomic masses, as was the case with O_2. A property of CO_2 is its ability to absorb infrared light, unlike O_2 or N_2. The intensity of infrared light once it has passed through the expired gas containing CO_2 gives a measure of the relative partial pressure of the CO_2 within the expired gas. Both methods are rapidly responding, and have been used in all types of gas concentration measurement systems.

Breath-by-breath systems are dependent on gas analysers with a very fast response time, such as the mass spectrometer. A response time is the time it takes an analyser to reach the true value from the value it started at. For example, if an analyser is measuring the concentration of O_2 in air, the analyser would be reading 20.93%. If the analyser is then suddenly presented with expired gas with a hypothetical O_2 concentration of 15.12%, the response time would be the time it takes the reading on the analyser to move from 20.93% to 15.12%. Usually, however, the 95% response time is required, which is the time taken for the analyser to reach 95% of the total change between 20.93% and 15.12%. This difference is 5.81%, and 95% of this value is 5.52%. By subtracting 5.52% from 20.93% we get the 95% value of 15.41%. Once the time is determined for the analyser to reach 15.41% from 20.93%, the 95% response time is known. The 95% response time is used in preference to the 100% response time because the 95% response time is usually significantly less.

If attempts are made to determine the **end-tidal** gas concentrations, it is necessary to have analysers that have an extremely rapid response time, since several samples must be made within a single breath. If one breath exhalation takes 1.5 seconds during heavy exercise, and 15 samples are needed during that time, it is necessary to have analysers that are sampling ten times per second. The response time of the analysers must also be less than one tenth of a second.

4 The effect of temperature on gas volumes

Even though the total number of gas molecules remains constant with changes in temperature, gas temperature is directly related to gas volume since an increase in temperature results in faster movement of gas molecules (i.e. Charles's law). Therefore, gas temperature should always be measured when determining respiratory variables. The temperature of a gas is determined in different ways depending on the system used. For off-line systems, the temperature is usually determined before the gas enters the volume meter, usually by a small temperature probe inserted into the gas flow. For on-line systems, the temperature of the gas is usually estimated based on the short time that the gas has been outside the body. Clearly, as the gas leaves the body the temperature will be the same as that of the body (i.e. 37°C).

5 The effect of ambient pressure on gas volumes

Again, even though the total number of gas molecules remains constant, the pressure acting on a gas is directly related to gas volume since an increase in pressure forces the molecules closer together (i.e. Boyle's law). Therefore, ambient pressure should always be measured when determining respiratory variables. Generally ambient pressure does not change rapidly, so a measurement at the start of an expirate collection period is adequate. If the weather conditions are changing rapidly, it may be necessary to make more frequent pressure measurements, perhaps one in the morning and one in the afternoon. Ambient pressure is normally measured with a mercury barometer, hence the common unit of ambient pressure, millimetres of mercury (mmHg). For further discussion of pressure see Chapter 5.

6 The effect of gas humidity on gas volumes and concentrations

Once again, even though the total number of gas molecules remains constant, the amount of water vapour (i.e. humidity) in a gas is directly related to gas volume since an increase in humidity adds an extra constituent to the gas. Therefore, humidity of the gas should always be measured when determining respiratory variables. Normally, gas expired from the human body (i.e. expirate) is saturated with water vapour. Evidence of this fact is seen as expirate cools in a Douglas bag, and condensation forms on the inside of the bag. The water vapour has come from the moist airways between the nose and mouth and the lungs. Inspirate normally contains some water vapour but is rarely saturated, which is why following exercise participants sometimes complain of a dry throat. The water in the saturated expirate has come from the participant's airways. The absolute amount of water vapour in a gas is a function of temperature. For example, gas at a temperature of 26°C can hold the equivalent of 25.2 mmHg of water vapour pressure whilst gas at a temperature of 39°C can hold 52.4 mmHg of water vapour pressure. In most physiology textbooks, tables are provided which indicate the water vapour pressure of a saturated gas for a given temperature (e.g. Robergs and Roberts 1996, appendix B, pp. 794–795). When the volume of gas is measured, water vapour is corrected for in the conversion from ATPS to either STPD or BTPS.

Gas humidity has an effect on the constituent gas concentrations in addition to overall gas volume. If it is assumed that water vapour is simply another gas in the expirate, the other constituent gas proportions will be distorted. When gas analysis techniques are used which measure a total amount of gas (e.g. mass spectrometry), water vapour is not a problem. However, most gas analysis techniques are based on partial pressures of the constituent gases (e.g. para-magnetic, fuel cell analysers), so distortion due to water vapour is a problem. To overcome this, when the concentration of gases in expirate is measured, rather than attempting to correct for the water vapour the gas should be dried prior to measurement. This is normally done with a condensing unit. Although the condensing unit will not remove all the water vapour, the important thing is that all gas is dried to the same degree.

Conclusion

The study of gases is underpinned by more than 300 years of scientific investigation. Gases do not occupy a fixed shape or volume unless enclosed and are easily expandable due to the weak attraction of the bonds between molecules. Gases are less dense than solids or liquids. The low density allows gas molecules to travel at high speeds. Calculation of individual gas molecule speeds is not possible, therefore an average velocity according to the kinetic theory of gases has been formulated. Three gas laws, known as Boyle's law, Charles's law and the pressure law, are particularly applicable to physiology. All three laws involve manipulating the variables of pressure, volume and temperature. A fourth variable, a mole, denoting the amount of a substance, is also represented in an equation known as the ideal gas law. Another law which has relevance to work in physiology laboratories is Dalton's law, which states the relationship between total pressure and the pressure of individual gases. Avogadro's law is also applicable as it states the relationship between different gases and volume. Under conditions of the same pressure, volume and temperature, gases will possess the same number of molecules.

The determination of respiratory variables in physiology is so common that the gas laws should form an integral part of a student's understanding about gas volumes. Since gas volumes are collected under different conditions of pressure and temperature, these variables must be carefully monitored. There are a number of different instruments which measure respiratory gases, ranging from standard off-line systems to fully automated on-line systems.

KEY POINTS

- Temperature, pressure and volume have significant impacts on the behaviour of gases.
- Gas molecules possess little molecular attraction or repulsion between one another.
- Boyle's law states that, under constant temperature conditions, pressure is inversely proportional to volume.
- Charles's law states that, under constant pressure conditions, volume is directly proportional to temperature.
- The pressure law states that, under constant volume conditions, temperature is directly proportional to pressure.
- The flow of a gas is dependent upon pressure gradients: a gas will move from a region of high pressure to one of low pressure.
- For a gas to flow from the environment into a liquid, three factors are directly linked to its flow: (a) pressure gradient; (b) solubility of the gas; (c) temperature.
- Dalton's law refers to the pressures of the individual gases which form a mixed gas: the total pressure of a mixed gas is the sum of the pressures of the individual gases.
- Diffusion, as applied to a mixed gas, is influenced by the total pressure but is also applicable to an individual gas and its partial pressure.
- Respiratory gas volume measurements are dependent on such factors as the temperature of the gas, the barometric pressure, and the relative humidity of the gas.
- The classic technique for the collection of respiratory gas is the Douglas bag, but there are now many automated systems available.

Bibliography

Robergs, R.A. and Roberts, S.O. (1996) *Exercise Physiology: Exercise, Performance, and Clinical Applications*. St Louis, MO: Mosby.

Further reading

Scratcherd, T. (1992) *Aids to Physiology*. Singapore: Churchill Livingstone/Longman.

3

LIQUIDS

AIMS OF THE CHAPTER

This chapter aims to provide an understanding of the scientific principles of liquids which are relevant to the theory which underpins sport and exercise. After reading this chapter you should be able to:

- state the properties of liquids;
- understand how liquids are affected by temperature and pressure;
- state the concept of buoyant force and Archimedes' principle;
- state Pascal's principle;
- understand the key concepts in relation to pressure in flowing liquids;
- understand the terms viscosity, laminar flow and turbulent flow;
- understand the terms moles, equivalents, per cent of solution and pH;
- apply the scientific principles of liquids to situations in sport and exercise.

Introduction

The term 'fluid' can be defined as 'a substance which flows', which, by definition, could include gases as well as liquids. There are many examples of sport and exercise activities taking place in a fluid environment, for example swimming, water aerobics, sailing and trampolining. This chapter will, however, refer to fluids only in the context of liquids (see Chapter 2 for discussion of gases), although there will be times when the laws/principles apply equally to gases. To appreciate the laws and principles as applied to gases it is advisable to read this chapter in conjunction with Chapter 2. In instances where laws/principles apply not only to a liquid but also to a gas, the text will refer to the dual application.

The majority of the earth's surface is covered by water. Water is also the main constituent within the human body and accounts for the largest proportion of total body mass. On average, the percentage of water in relation to total body mass is 60% for males and 51% for females. There are of course many other liquids in the body such as synovial fluid, cerebrospinal fluid, mucus and saliva, all of which are water-based solutions. The chemical formula for water is H_2O. The hydrogen bond with oxygen is the result of a weak interaction between a positive region of the hydrogen and the negative region of an oxygen atom. It is the hydrogen bonding in the water molecule that is responsible for water **surface tension** (defined as the cohesive force between liquid molecules which creates a surface film). As water is so important within the human body, it is worth noting some of its characteristics:

- it is liquid at standard temperature;
- it has a large heat of fusion;
- it has a large heat capacity;
- it has a large heat of vaporisation;
- its **density** at 4°C is 1 g·cm^{-3}.

Scientific principles of liquids

The molecular structure of gases and solids is covered in Chapters 2 and 4 respectively. Liquids, like solids, have a very strong bond holding their atoms together. This strong bonding is termed **cohesion**, which implies that there is an attraction of like molecules for one another. The cohesiveness of a liquid has implications for such concepts as surface tension and **viscosity** (defined as resistance to flow). Liquids, unlike solids, are able to move more freely and hence will take the shape of any container in which they are placed. The molecules within a liquid move more slowly than those of a gas. The molecules do, however, continue to vibrate and collide. There is also less space between the molecules of a liquid so they occupy much less volume than the same number of molecules of a gas. These rapidly moving molecules cannot be observed with the human eye. However, if a very fine particle of dust is dropped into a beaker of liquid, its random movement through the liquid can be observed using a microscope. The movements of the particles are random, because the fast-moving molecules are continually colliding. Indeed, a botanist, Robert Brown (1773–1858) observed this type of movement, hence its name, Brownian motion. As the molecules rapidly move around inside the beaker, some of the fastest-moving molecules free themselves from the attractive forces of nearby molecules and escape from the liquid into the gas above it. This process is known as evaporation, and it is those molecules with the greatest energy that escape. The overall effect is a reduction in the average energy of the remaining molecules, resulting in a decrease in the temperature of the liquid.

A further property of a liquid is its incompressibility: it is extremely difficult to force a liquid to occupy a smaller volume than it is already occupying. Therefore, two distinguishing features of liquids are:

- At a particular temperature, a liquid has a fixed volume but will take the shape of any container.
- Pressure applied to a liquid will not change the volume (unlike a gas).

Hence, the state of a liquid will vary according to the surrounding temperature and pressure. Clearly, the addition of heat energy will change a liquid to a gas, a process known as vaporising. Cooling a gas into a liquid is known as condensation. The changing of a liquid to a solid is a process known as freezing.

Pressure in stationary liquids

The definition of **pressure** is given as force per unit area (the SI unit is the pascal, Pa), which is equal to a pressure of 1 newton per square metre (1 N·m^{-2}). The definition of force and pressure and the units of measurement of force and pressure are covered in Chapters 1 and 5. The pressure in a stationary liquid is a function of the depth of the liquid and its density. If a swimmer dives into water, the deeper the dive, the greater the pressure upon the body. This pressure is directly proportional to the depth of the liquid and the density of the liquid (see Figure 3.1). The pressure exerted by the liquid is a consequence of the contact it has with the container and also of the fact that the liquid has weight.

The relationship between the depth of the liquid and the pressure of the liquid can be described as follows:

$$P = \rho g h, \tag{3.1}$$

where P is pressure of the liquid, ρ is density, g is acceleration due to gravity, and h is depth of the liquid.

For example, if a container held a liquid with a depth of 5 m and the liquid within the container had a density of 1 g·cm^{-3} (1 g·cm^{-3} = 1000 kg·m^{-3}), the pressure in the liquid would be:

$$P = \rho g h$$
$$= 1000 \text{ kg·m}^{-3} \times 9.8 \text{ m·s}^{-2} \times 5 \text{ m}$$
$$= 49\,000 \text{ N·m}^{-2} = 49\,000 \text{ Pa}$$
$$= 49.0 \text{ kPa}.$$

(Note that the density measure has been multiplied up to ensure SI units in the calculation.) Note that the pressure at any point within the liquid acts in all directions.

A simple example of pressure exerted on a stationary liquid is seen in the barometers used in laboratories to record atmospheric pressure (see Figure 3.2). In a barometer, one of the ends of the tube is closed and evacuated, creating a vacuum. In this arrangement, the absolute pressure can be measured. In laboratory barometers mercury is typically used. As mercury has a density of 13.6 g·cm^3, it will require 13.6 times the pressure to push mercury to a particular height as it would to push water to the same height. In laboratory experiments that involve collecting respiratory gases (oxygen and carbon dioxide), it is important to

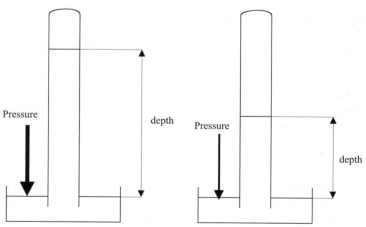

Figure 3.1 The relationship between depth and pressure in a stationary liquid

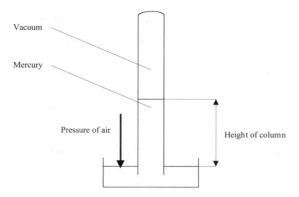

Figure 3.2 The principle of a barometer

know the barometric pressure under which the gases are collected (see Chapter 2). Typically, at sea level the barometric pressure is 101 kPa. Using a mercury barometer and Equation 3.1, if a column of mercury were raised 760 mm high, the atmospheric pressure could be calculated as follows:

$P = \rho g h$

$\quad = 13\ 600\ \text{kg·m}^{-3} \times 9.8\ \text{m·s}^{-2} \times 0.76\ \text{m}$

$\quad = 101\ 396.2\ \text{N·m}^{-2} = 101\ 396.2\ \text{Pa}$

$\quad = 101\ \text{kPa}.$

(Note that because there is some rounding of the number it does not add up exactly to 760 mm of mercury raised.)

For various reasons, although kPa is the correct SI unit for pressure another unit is more commonly used in the literature (see Action point 1). At sea level an atmospheric pressure of 101 kPa, which is also known as standard pressure or 1 atmosphere, is defined as the pressure exerted at the foot of a column of mercury with a specific density of 13.6 g·cm^{-3} and at a specific gravity value sufficient to raise the mercury 760 mm high.

ACTION POINTS

1   Although it is possible to measure pressure in mm H$_2$O, what is the most common unit used? Consider also what units weather forecasters usually use for their pressure readings.
2   Consider how you could use some tubing and water to measure the force exerted by the respiratory muscles during either inspiration or expiration.
3   Define pressure at a point in a liquid.
4   Since most of the body is composed of liquids (mostly water) why is there no danger that the body will be crushed or compressed by increasing pressures found with ever increasing depths from underwater diving?

Buoyancy force and Archimedes' principle

The effects of buoyancy are clearly seen when an object floats in water. The object appears to lose its weight and be supported within the water. The effect of buoyancy is due to an upward lift or upthrust of water, which is in turn due to the pressure exerted by the water being greater on the lower parts of the submerged object than on the top of the object (remember pressure increases with depth). This situation also applies to gases, as can be seen when a hot air balloon floats into the air. Hence:

If the buoyant force is equal to the weight of the object it will float.

If the buoyant force is less than the weight of the object it will sink.

Therefore, the buoyant force can be described as the difference between the liquid pressure above and below the object. The law that describes this phenomenon was discovered by Archimedes (287–212 BC) and is known as Archimedes' principle:

The buoyant force on a completely or partially submerged object is equal to the weight of the fluid displaced.

Note the use of the word 'fluid': this principle applies equally to gases as to liquids.

As the upthrust on the object will be equal to the weight of the immersed object, the weight of the displaced liquid can be measured. Archimedes used this principle to weigh objects in air and in water: the difference between the two is known as the buoyancy force. In this way the volume of an object, and also its density (by dividing mass by volume), can be determined.

It is often common practice in the study of sport and exercise to estimate percentage body fat. In order to calculate this, underwater weighing is often used to determine the density of the human body by using the principles of buoyancy force and Archimedes' principle. The density of a human body can then be converted, using standard equations, into a percentage body fat score. If a human body weighing 75 kg is submerged in water and displaces 69 kg of water, the weight under water must have been 6 kg (75 kg – 6 kg = 69 kg). If the density of the water is 1 g·cm^{-3} (the density of water at 4°C), then the specific volume of the body will be 0.069 m^3 (69 kg/1000 kg·m^{-3}). Next, the body density is calculated as 75 kg/0.069 m^3 = 1086.9 kg·m^{-3} = 1.087 g·cm^{-3}. Since the water is only 0.92 times as dense as the body, the resulting buoyancy force of 69 kg will not overcome the weight of the body and therefore it will sink.

If the same example is examined again, but this time the human body has a submerged weight of only 3 kg, the resulting displacement of water will be 72 kg. The volume of the body is 0.072 m^3, which results in a body density of 1.041g·cm^{-3}. Similarly, the resulting buoyancy force will not overcome the weight of the body and it will sink. Although there are a number of other factors which must be considered during underwater weighing, for example the amount of air remaining in the gastrointestinal tract and the residual volume left in the lungs after maximally exhaling during the submersion, the procedure is common practice in sport and exercise science.

The above observation is known as the 'principle of flotation':

A floating object will displace its own weight of fluid.

The Archimedes principle is routinely employed in hydrodensitometry, or underwater weighing, where a human body is immersed in a tank of water and the weight of the submerged body is recorded in order to calculate the body's density. Another instrument which uses the same flotation principle is a hydrometer, used to determine the specific gravity of a liquid (see Figure 3.3). A hydrometer is floated in the liquid and a reading may be made off the scale inside the stem, level with the liquid's surface. In certain occupations, such as nursing, a urinometer is used to measure the specific gravity of urine, as some diseases alter this.

Pascal's principle

Pascal's principle refers to the transmission of pressure to all parts of a liquid:

Any change of pressure in an enclosed fluid is transmitted to all parts of the fluid.

Earlier in the chapter it was shown that the density of a liquid multiplied by the depth determined static pressure within the liquid, and that the exerted pressure was equal in all directions. In order to calculate the total pressure on the liquid, any other external pressures must also be taken into consideration. For example, the atmosphere exerts a pressure of approximately 101.3 kPa (760 mmHg) at sea level. This external pressure is transmitted throughout the whole of the liquid. Pascal's principle is applied in such devices as hydraulic presses: a small amount of pressure is applied to an enclosed fluid, and because of its incompressibility and transmission of the pressure, very large forces can be generated. Other examples of such devices are hydraulic lifts, car brake systems, cutters used by the emergency services after accidents, and even barbers' chairs.

Pressure in flowing liquids

The previous examples demonstrate that the pressure within a stationary liquid is a direct function of the depth and density of the liquid. Likewise, as there is no flow in a stationary

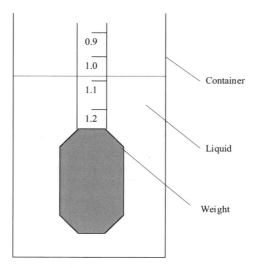

Figure 3.3 A hydrometer, an instrument used for measuring the specific gravity of a liquid

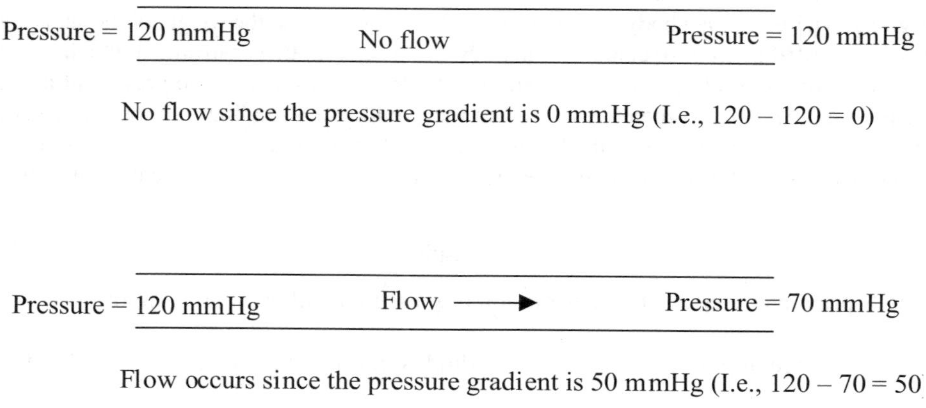

Pressure = 120 mmHg	No flow	Pressure = 120 mmHg

No flow since the pressure gradient is 0 mmHg (I.e., 120 – 120 = 0)

Pressure = 120 mmHg	Flow ⟶	Pressure = 70 mmHg

Flow occurs since the pressure gradient is 50 mmHg (I.e., 120 – 70 = 50

Figure 3.4 The relationship between pressure in a flowing liquid and the pressure gradient

liquid, there is no change in pressure. The pressure is therefore equal in all directions, as noted above. However, pressure within a flowing liquid means that further variables need to be considered. To understand the principles of pressure in flowing liquids it is important to consider four points:

- For a liquid to flow through a tube, there must be a drop in pressure (see Figure 3.4).
- The pressure will be lower at the exit of the tube than at the entrance to the tube.
- If there are equally spaced horizontal tubes placed in succession to one another the amount of pressure drop will be the same.
- It is conventional to use the term **pressure gradient** rather than pressure changes to describe the pressure in flowing liquids.

The pressure gradient can be defined as the pressure drop per unit length:

$$\text{Pressure gradient} = \text{Pressure drop/Length} \tag{3.2}$$
$$= P_1 - P_2/L,$$

where P_1 is pressure at point 1, P_2 is pressure at point 2, and L is the length of the tube.

As can be seen from Equation 3.2, the variable L (length) is introduced as the denominator of the pressure drop. Several other variables affect the rate of flow through a tube (see Figure 3.5). The volume flow rate (F) is the volume of flow per unit time and is dependent not only upon the drop in pressure but also upon the resistance to the flow within the tube. The volume flow rate may be calculated as follows:

$$F = P_1 - P_2/R, \tag{3.3}$$

where F is flow rate and R is resistance to flow.

Figure 3.5 shows that the pressure drop is characterised as representing losses of energy,

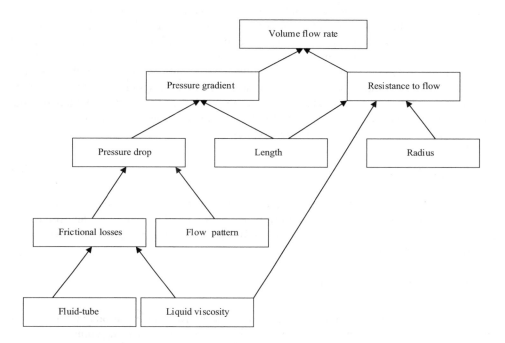

Figure 3.5 Schematic showing the variables that affect the rate of flow of a liquid through a tube

principally through frictional losses of the tubing, the fluid or the pattern of liquid flow. We will consider each of these in turn.

When the wall of a piece of tubing comes into contact with the fluid flowing through it, the friction will depend on the material of the wall. For example, rubber tubing causes greater **friction** between the fluid and the wall than plastic tubing, and for this reason the latter is more frequently used nowadays.

There can also be friction within the liquid itself. The force that opposes flow within a liquid is known as viscosity (symbol η). The viscosity of a liquid is due to the collective cohesive forces of the molecules in the liquid. The stronger the cohesive force, the greater is the resistance to flow. For example, human blood can be up to five times as viscous as water. As blood becomes more viscous more pressure will be required for it to flow at the same rate, which will naturally have consequences for the amount of stress placed on the heart. Therefore, providing the pressure gradient is constant, flow rate is inversely proportional to viscosity.

Loss of energy is also related to the type of flow pattern, which can be categorised as either **laminar** or **turbulent**. Laminar flow represents the minimum loss of energy and is representative of a tube which might be straight, and have smooth walls, so that the flow will tend to be in smooth layers. In fact, if it were possible to look at the flow of the liquid through smooth tubes, it would be seen that the liquid next to the walls is at rest whilst the maximum speed of the liquid is found at the centre of the vessel. As some parts of the liquid are at rest while other parts are travelling quickly, the mean speed of flow is approximately half the maximum speed found at the centre. Problems in fluid dynamics occur when the speed of the flow exceeds a critical value or if the liquid collides with an obstruction. Both of these situations create eddies, which results in the laminar flow being transformed into turbulent flow. Although there are a number of equations for calculating pressure gradients and flow rate, accurate prediction of the dynamics of turbulent flow has proved very difficult. This area continues to be the subject of extensive study by scientists across the world.

The study of liquids through cylindrical tubes was greatly influenced by the work of a French scientist, Poiseuille, who in 1844 investigated steady flows of liquid through a pipe. Poiseuille was interested in the relationships of steady flow so he could apply them to blood flow through veins and arteries. Poiseuille devised an equation to gauge the resistance to flow of a pure liquid under laminar flow conditions:

$$R = 8\eta L/\pi r^4, \tag{3.4}$$

where η is the viscosity of the fluid, L is the length of the tube, and r is the radius of the tube.

Equation 3.4 shows the resistance to flow. However, for flow rate to be determined, the pressure gradient needs to be introduced. Therefore, by substituting Equation 3.4 into Equation 3.3, Equation 3.5 is derived:

$$F = (P_1 - P_2)/(8\eta L/\pi r^4) = (\pi(P_1 - P_2)r^4)/8\eta L. \tag{3.5}$$

Equation 3.5 is known as Poiseuille's law. It can be seen from this equation that the radius is raised to the fourth power. This factor is an important determinate in volume flow rate calculations. In fact, the equation and the resulting calculations show that volume flow rate depends more on the radius of the tube than on the fluid pressure.

Although water as a pure liquid follows Poiseuille's law quite well, some liquids like blood do not behave precisely as expected. This is thought to be due to the many different types of substances that are found in blood in solution or suspension. For example, red blood cells may accumulate in the faster axial part of the flow, hence there are fewer of these cells to create friction against the vessel walls. The practicalities of this law will be discussed later, in the section on applications to sport and exercise.

ACTION POINTS

1   Why might it be suggested that somebody weighs less under water when clearly their mass has not changed?
2   If the force exerted on a surface which is in contact with a stationary liquid is not perpendicular to the surface at all points, what happens to the liquid?
3   Why would a fluid such as saline (salt water) have a lower viscosity than blood?
4   Determine a volume flow rate using Poiseuille's equation and then individually double one parameter at a time to examine the effects on the flow rate.

Bernoulli's principle

Another consideration relating to pressure and velocity in a moving fluid is Bernoulli's principle. Daniel Bernoulli (1700–1782) was an Italian scientist who found that the pressure in a flowing liquid is at its lowest where the speed of flow is greatest. The principle states that faster-flowing fluids exert lower pressures than slower-flowing fluids. Note the use of the word 'fluid': once again this principle applies equally to gases as to liquids. The equation for this principle can be expressed as follows:

$$P + \rho gh + \tfrac{1}{2} \rho v^2 = \text{constant},\tag{3.6}$$

where P is pressure, ρ is density of the fluid, g is gravity, h is height, and v is mean fluid velocity across the section.

The relationship between the variables in Equation 3.6 shows that an increase in speed is directly related to a drop in pressure. For example, for a liquid in a tube to flow there must be a pressure drop. If a tube is reduced in diameter in its middle section, the liquid would flow more quickly through the constricted area in order to transport the same volume of fluid in a given time. Bernoulli's principle can be applied to gases under certain conditions, including air travelling at moderate speeds. However, the principle is based on an incompressible fluid and, as gases are compressible, caution should be exercised if applying the principle to them.

Solutes, solutions and concentration

As already mentioned, a major component of the human body is water. Substances such as starch which enter the body will ultimately dissolve into watery cells. The substances which dissolve are known as **solutes** and the liquids into which they dissolve are known as **solvents**; the combination of a solute and a solvent results in a **solution**. If a solvent is described as being highly soluble, this means that molecules are easily dissolved in it, whereas if a solvent possesses a low solubility, this means that molecules are not easily dissolved in it.

Solutions are often described as having a **concentration**, and this relates to the amount of solute per unit volume of the solution. In physiology, there are a number of different ways in which the concentration of a solution may be expressed. These include:

- moles per unit volume ($mol \cdot L^{-1}$);
- equivalents (equiv.);
- weight/volume ($g \cdot L^{-1}$);
- per cent of solution (%).

Moles

A mole can be defined as 6.02×10^{23} atoms, ions or molecules of a substance. Therefore, one mole of one substance has exactly the same number of atoms as one mole of any other substance. The weight of a mole of a substance is, however, different for different substances. Therefore, we can use the term 'molecular weight', which is the equivalent of the combined atomic mass of the substance expressed in grams. For example, glucose has a chemical formula of $C_6H_{12}O_6$; the atomic mass of carbon is 12 g, that of hydrogen 1 g and that of oxygen 16 g (check the periodic table to confirm the atomic masses), therefore the resulting molecular weight is 180 g ($6 \, C \times 12 \, g + 12 \, H \times 1 \, g + 6 \, O \times 16 \, g$).

In biology it is often common to refer to concentration as **molarity**. This is the number of moles of solute in 1 litre of solution, and is often signified as $mol \cdot L^{-1}$ or M. Therefore, a 1 M solution of glucose is made by dissolving 1 mole or 180 g of glucose into 1 litre of water. A derivative of a mole is the millimole (mmol), which is 1/1000th of a mole. The mmol is commonly used because often solutions are very dilute.

Equivalents

Another method of expressing concentration is to refer to the concentration of ions, and this is represented as equivalents per litre. An equivalent is the sum of the molarity of the ion multiplied by the number of charges the ion carries. A sodium ion, often written as Na^+, has a single charge of +1 and is therefore one equivalent per mole. The same rule is followed for negatively charged ions: for example, the phosphate ion, HPO_4^{2-}, has two equivalents per mole. As with mmol, a milliequivalent (mequiv.) is 1/1000th of an equivalent.

Weight by volume and per cent of solution

On many occasions it is not possible to use solutes by the mole and so the more conventional measure of weight is instead used. The solute concentration can be expressed as a percentage of the total solution. A 25% solution means that there are 25 parts of solute per 100 parts of the total solution. However, if the solute represents a solid, then weight per unit volume must be used to represent the percentage solution. Hence, an 8% glucose solution would be made by weighing out 8 g and adding enough water to make a total solution volume of 100 mL. If the solute is already a liquid, the solution uses a volume per unit volume measure.

Hydrogen ions

One of the most important solutes in the human body is the hydrogen ion, H^+. As indicated by the symbol, the hydrogen ion has a positive charge due to the loss of an electron, and therefore consists of only a single proton. These positively charged ions are called cations. In the same manner an ion symbolised as negatively charged, for example a chloride ion, denoted by Cl^-, has an extra electron. Negatively charged ions are called anions. Thus a calcium ion (Ca^{2+}) has lost two electrons and a sulphate ion (SO_4^{2-}) has gained two electrons.

The concentration of H^+ in the human body has important consequences for physiological function, particularly in relation to acidity. Hydrogen ions are constantly produced in the body via H_2O when it separates into H^+ and OH^- or from ionised molecules that release H^+ when dissolved in water. A molecule donating a hydrogen ion to a solution is known as an acid. Molecules that decrease the concentration of H^+ in a solution, by combining with the free H^+, are known as bases. The H^+ concentration within the body is measured as pH, which stands for power of hydrogen. The pH is calculated as the negative log of the hydrogen ion concentration:

$$pH = -\log[H^+].\tag{3.7}$$

Note the use of the square brackets [ ], which symbolise concentration.

Equation 3.7 can be rewritten as:

$$pH = \log(1/[H^+]).\tag{3.8}$$

Equation 3.8 states that the pH is inversely related to the concentration of hydrogen ions: as pH goes up, H⁺ concentration goes down. Figure 3.6 shows the pH scale, which runs from 0 to 14. Pure water has a hydrogen ion concentration equal to 1×10^{-7} M and is therefore classified as having a pH value of 7, or 'neutral'. Solutions that have gained H⁺ from an acid will have a higher [H⁺]. Thus acidic substances like vinegar have a pH score of 3 and an H⁺ concentration of 1×10^{-3} M. What is not obvious from Figure 3.6 is that the pH scale is actually based on a logarithmic interval, such that a 1 unit increase, for example from 6 to 7, will represent a tenfold decrease in the H⁺ concentration (remember the closer pH moves towards 14 the more alkaline is the solution becoming). Similarly, if the pH value moves from 7 to 4, there has been a 100-fold (10×10) increase in the H⁺ concentration. During moderate or heavy exercise, increasing acidity within cells and in the blood have to be dealt with by the body. Sport and exercise physiologists have for a long time suspected acidity as a possible cause of muscle fatigue during exercise.

APPLICATION OF SCIENCE TO EXERCISE AND SPORT

The preceding discussion on the principles and laws of liquids has given some examples of their application in a sport and exercise context. Other useful examples to note at this stage are:

- working out solutions for sports drinks, and how concentrated the glucose should be in relation to water (see section on solutes, solutions and concentration);
- the use of a water manometer to examine lung function (see section on pressure in stationary liquids);
- the relationship between blood lactate concentration ($[La^-]_B$) and the H⁺ concentration ($[H^+]$) in the body. In exercise physiology $[La^-]_B$ is often used as a measure of how hard an individual is working during a specific exercise test or sports event (see section on pH).

Blood pressure

Blood pressure is often determined during routine medical examinations, and used in a sport and exercise context prior to a fitness test. Blood pressure could also be taken during a laboratory experiment to investigate the blood pressure response to exercise. An athlete or patient who is diagnosed as being hypertensive (i.e. having high blood pressure) would have to undergo a more thorough medical examination before being allowed to undertake an exercise test.

Blood pressure is a reflection of the driving pressure created by the ventricular contraction of the heart (i.e. in systole) and the residual pressure in the circulatory system (i.e. in diastole). As the blood which leaves the left ventricle is under the greatest pressure, systolic blood pressure is a reflection of the highest pressure exerted by the left ventricle during a cardiac cycle. Diastolic blood pressure is the lowest pressure recorded during a cardiac cycle, while the ventricles are relaxing. Blood pressure is very difficult to measure directly from the ventricles, hence it is often estimated indirectly from the radial artery of the arm. This method is assumed to be indicative of the driving pressure for blood flow. During ventricular

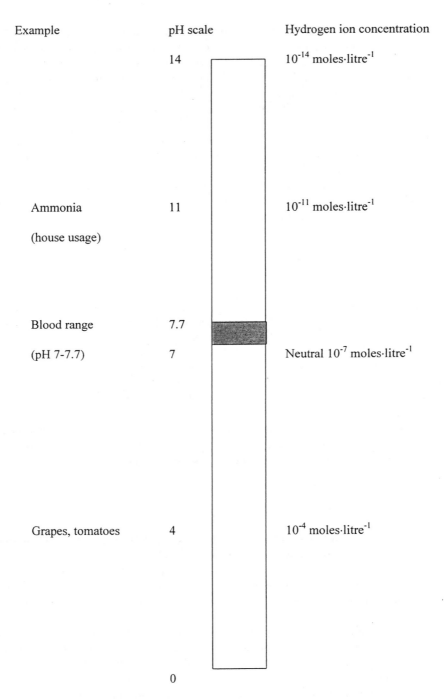

Example	pH scale	Hydrogen ion concentration
	14	10^{-14} moles·litre^{-1}
Ammonia (house usage)	11	10^{-11} moles·litre^{-1}
Blood range (pH 7-7.7)	7.7 7	Neutral 10^{-7} moles·litre^{-1}
Grapes, tomatoes	4	10^{-4} moles·litre^{-1}
	0	

Figure 3.6 The pH scale and related values

contraction, the reading is on average 120 mmHg and this falls to 80 mmHg during ventricular relaxation. One of the important considerations for blood pressure is the ability of the aorta and arterioles to expand and store the energy in their elastic walls. This elastic properties of the walls, which enables them to expand and store energy, results in a blood flow to the arterial side that is pulsatile (wave-like).

The pulse that a person can feel in their arm or through their chest is a result of the rapid contraction of the ventricles forcing blood through the aorta. The result of this pulse is to create a pressure wave that transmits itself through the blood and travels through the arteries and arterioles. This pressure wave is actually travelling faster than the blood, in fact about ten times as fast. Therefore, the pulse that is felt in the arm occurs after the contraction of the ventricle that created the pressure wave. Depending upon the distance the wave has travelled and the resulting friction, the size of the pulse wave will decrease and the flow of the blood will become smoother rather than pulsatile.

In physiology there is a parameter called pulse pressure which is an index of the amplitude of the pulse pressure wave. Pulse pressure is calculated as follows:

Pulse pressure = Systolic pressure – Diastolic pressure,   (3.9)

e.g. 40 = 125 – 85.

Owing to the distance the blood has had to travel to pass through the veins, the pressure decreases dramatically. The veins below the heart are therefore known as low pressure capacitors (remember that the pulse wave disappeared because of friction). The blood in the veins below the level of the heart must be returned back in circulatory fashion to the heart. Although the pressure does not increase as the blood travels towards the right atrium, the blood does get help from another type of pump, the skeletal muscle pump. The action of muscle contractions in the calf or upper leg results in the constriction of the veins and the blood is squeezed upwards. To assist this flow, which is known as venous flow, veins have valves that prevent the blood flowing backwards, similar to the valves in the heart. This ensures that blood is forced upwards. One of the main reasons why soldiers on parade faint or collapse is because of venous pooling of the blood in the lower legs. As there is very little muscle contraction of the lower legs during the long hours some soldiers have to stand, less blood is forced upward to the heart and more blood is left pooling in the veins. This results in less blood circulating and transporting oxygen around the body, particularly to the brain, and when a critical point is reached the body's natural response is to faint, and the victim is brought crashing to the ground. As a result the pooled blood in the legs will begin to re-circulate and, providing there are no other serious injuries from the fall, the soldier should quickly regain consciousness.

As described previously, arterial blood pressure is estimated from the radial artery in the arm. To measure blood pressure directly, an indwelling catheter would have to be placed directly into an artery. The sphygmomanometer (from sphygmus – pulse, and manometer – instrument for measuring fluid pressure) uses an inflatable cuff device strapped around the upper arm (usually the left). The cuff is inflated to a higher pressure than the systolic pressure and as a result of this the blood flow into the lower arm is stopped. As the cuff is slowly deflated the cuff pressure approaches the systolic pressure. Eventually the cuff pressure drops just below the systolic pressure and blood begins to squeeze through the constricted artery. Blood flow is turbulent at this point, and causes what is known as a Korotkoff sound.

This sound can be heard through a stethoscope placed just below the cuff. For the sound to be heard the artery has to be compressed, and each Korotkoff sound is heard with each heart beat. It is at this point that the reading on the sphygmomanometer is recorded. As the cuff pressure continues to fall, the pressure exerted on the artery becomes less and less until eventually the artery is compressed no longer and the Korotkoff sounds disappear, since blood flow is no longer turbulent. It is at this point that the second reading, the diastolic pressure, is recorded. The two readings which are taken, i.e. of systolic/diastolic pressure, are 120/80 mmHg in a person with normal blood pressure. For a person to be considered hypertensive (having high blood pressure) a blood pressure reading of 140/90 would have to be found several times, as there is a wide variation within a single individual from one moment to the next.

In addition to pulse pressure, mean arterial pressure (MAP) is used to describe blood pressure. The arterial pressure is a wave, and therefore pulsatile, so a single value is thought to be useful to indicate the driving pressure. To calculate the MAP, which typically applies only to a heart rate range between 60 and 80 beats per minute, the following formula is used:

$$\text{MAP} = \text{Diastolic pressure} + \tfrac{1}{3}(\text{Systolic pressure} - \text{Diastolic pressure}) \qquad (3.10)$$

$$= 82 + \tfrac{1}{3}(125 - 82)$$

$$= 96 \text{ mmHg.}$$

The calculation of MAP gives a figure that is closer to the diastolic figure than to the systolic one, and this is because at rest the heart is in diastole twice as long as it is in systole. This is why the formula will work only for a resting heart rate value. For example, if the heart rate increases then the amount of time spent in diastole (relaxation period of the heart's ventricles) decreases and therefore the contribution of systole (contraction phase of the heart's ventricles) becomes relatively more important for the MAP.

MAP is a major determinant of blood flow and is influenced by four factors:

- resistance to blood flow within the circulatory system;
- cardiac output (Q);
- relative distribution of blood between arterial and venous vessels;
- blood volume.

The site of greatest resistance is the arterioles, which contribute over 60% of the total resistance to flow in the circulatory system. The resistance is principally due to the amount of smooth muscle possessed by the arterioles, and it is this smooth muscle that enables the arterioles to constrict and hence decrease flow or dilate and hence increase flow. As has already been noted, a very small change in radius creates a large change in the resistance (see Equation 3.4). By applying the fundamentals of pressure, flow and resistance, Equation 3.4 may be applied to physiological responses such as blood flow and pressure. For further development of knowledge in this area, topics such as blood volume and how the blood is distributed throughout the body should be examined.

ACTION POINT

1   Consider the scenario of a student who was late for a practical on blood pressure and had to rush to get to the laboratory. What would be the effect of the increased heart rate on his or her blood pressure?

Water skiing

Bernoulli's principle can be applied to a water skier who is about to be pulled from a crouched position into the upright position and begin skiing behind a speed boat. At first, as the increase in velocity begins to pull the skier upright, there are considerable differences in the velocity of flow of water beneath the skis compared with that above them. As the velocity of the skier begins to increase, the water underneath the skis will be at a lower velocity than that travelling over the top of them. Hence, at the bottom of the skis, the slower water will result in a higher pressure, and at the top of the skis, the faster-travelling water will result in a lower pressure. The relative motion between the skis and the water results in a force perpendicular to the flow. Because beneath the skis the pressure is increased, and above the skis the pressure is reduced, an upwards force termed 'lift' is created. Figure 3.7 shows application of the principle to water skiing. The lift is sufficient to allow the water skier to move into the upright position and to support their weight as they ski behind the boat. The principle works in exactly the same manner for hydrofoils, aircraft wings and propellers (aerofoils) except that in these cases it is air, rather than water, flowing over an object that creates the lift.

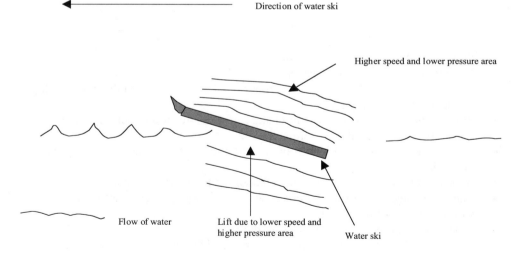

Figure 3.7 Application of Bernoulli's principle to water skiing

Conclusion

Although the behaviour of some aspects of liquids is difficult to predict, an extensive amount of research has been done on them. In the third century BC, work by Archimedes established a principle which stated that the buoyant force on a submerged or partially submerged object was equal to the weight of the fluid displaced. If the volume of fluid could be calculated, and if an object's mass were known, the density of an object could be determined. These principles have been put into practice to determine such variables as body fat, measured using the technique known as hydrodensitometry (underwater weighing).

A property of liquids is their fixed volume. It is known that in liquids the bonding between molecules is strong, and this accounts for their incompressibility. This incompressibility under pressure has been verified from work by Pascal, who found that pressure within an enclosed liquid was transmitted equally. Pascal's principle has led to devices such as hydraulic presses and braking systems. Pressure affects not only stationary liquids but also flowing ones. Many physiological responses in the body involve flowing liquids. For a liquid to flow there must be a drop in pressure, and this is often referred to as the pressure gradient. A number of other factors influence flowing liquids, such as length of the vessel through which the liquid flows, the resistance offered by the vessel and the internal resistance of the liquid itself. Such scientific facts allow investigation of conditions such as hypertension (high blood pressure).

Besides the issues discussed in this chapter, there is a whole branch of engineering devoted to the study of the mechanical properties of fluids, known as hydraulics. A subdivision of hydraulics is hydrostatics, which deals with the study of liquids at rest. Another subdivision is hydrokinetics, which deals with liquids in motion and is particularly concerned with friction and turbulence generated by flowing liquids.

KEY POINTS

- Pressure is defined as force per unit area.
- Liquids are essentially incompressible.
- In a stationary liquid the pressure is proportional to the density of the liquid multiplied by the height of the liquid.
- Pascal's principle states that any change in pressure in an enclosed liquid will be transmitted evenly to all parts of the liquid.
- Any submerged object will have an upward buoyant force equal to the weight of the liquid displaced.
- If the density of an object is less than that of the liquid, it will float.
- The percentage of the floating object which is submerged gives an indication of the specific gravity of the liquid.
- When a liquid flows there is a pressure drop.
- Poiseuille's law states that the pressure drop divided by the resistance is equal to the flow rate.
- The resistance to flow is most strongly influenced by the radius of the tube through which the liquid flows.
- Although blood is not an ideal liquid, unlike water, Poiseuille's law can be used to describe some of the flow dynamics of it.

- It is more accurate to use the term 'pressure gradient' than 'pressure change'. If there is no pressure gradient there is no flow of liquid.
- Flow rate is most affected by the radius of the vessel through which the liquid travels.
- Obstructions to the flow of a liquid are likely to cause disturbances in laminar flow and possibly lead to turbulent flow.
- Bernoulli's principle states that in an incompressible fluid the sum of its pressure and square of velocity is always constant.
- A solute plus a solvent equals a solution. The solute is the dissolved substance and the solvent is the liquid into which the solute was dissolved.
- A concentration describes the ratio of the solute and the solution. It can be expressed in moles (mol), equivalents (equiv.), weight per unit volume (g·L^{-1}) or per cent of solution (%).
- Blood pressure represents the driving force of the ventricles during systole and the residual pressure within the circulatory system during diastole.
- The resistance to blood flow is an important factor in the physiology of circulation.

Further reading

Duncan, T. (1994) *Advanced Physics*, 4th edition. London: John Murray Publishers.

Green, J.H. (1966) *An Introduction to Human Physiology*. London: Oxford University Press.

Nave, C.R. and Nave, B.C. (1985) *Physics for the Health Sciences*, 3rd edition. Philadelphia, PA: W.B. Saunders.

Rose, S. (1991) *The Chemistry of Life*. London: Penguin Books.

Silverthorn, D.U. (1998) *Human Physiology. An Integrated Approach*. Englewood Cliffs, NJ: Prentice-Hall.

Uvarov, E.B. and Isaacs, A. (1986) *The Penguin Dictionary of Science*, 6th edition. London: Penguin Group.

4

SOLIDS

AIMS OF THE CHAPTER

This chapter aims to provide an understanding of the scientific principles of solids which are relevant to the theory which underpins sport and exercise. After reading this chapter you should be able to:

- state the properties of solids;
- appreciate the term internal energy;
- understand how solids are affected by temperature;
- understand the terms conduction, thermal conductivity and radiation;
- apply the scientific principles of solids to situations in sport and exercise.

Introduction

'Solid' is defined as 'the physical state of matter in which the constituent molecules, atoms or ions have no translatory motion although they vibrate about the fixed positions that they occupy in a crystal lattice' (Uvarov and Isaacs 1986: 376). A solid, such as metal, possesses the property of stiffness or rigidity. This is due to the atoms possessing order and being held in approximately fixed positions. Molecules forming a solid are typically arranged in an orderly fashion known as a lattice. Most solids will thus keep their shape and, unlike liquids, will not take the shape of a container that they are placed into. Although the human eye sees a solid as a fixed state of matter, on a molecular scale the picture is very different. The atoms of a solid are in fact moving continuously and rapidly. The molecules in solids cannot move as quickly as those in gases because in solids the movement of the molecules is limited by the attraction between them. This means that in a solid there is limited space in which the molecules can move, unlike in gases. However, there is a large amount of kinetic energy in a solid.

Scientific principles of solids

Solids can possess all or some of the following characteristics:

- thermal conductivity;
- electrical conductivity;
- mechanical conductivity.

In terms of mechanical properties, solids can have four different characteristics: (a) strength, which refers to how great an applied force a material can endure before breaking; (b) stiffness, which refers to the opposition a material presents against being distorted by having its shape and size altered; (c) ductility, which is the ability of a material to be hammered, rolled and stretched into a useful shape; (d) toughness, which relates to a material which is not brittle. Some solids, such as steel, possess all these properties whereas others possess only some. Glass, for example, is strong and stiff but not ductile or tough.

As stated in the introduction, the atoms of a solid are alternating between attracting and repelling one another whilst vibrating around an equilibrium point. This movement energy is known as **internal energy** and is due to both the kinetic and the potential energy of the atoms in the solid. In a solid there is approximately equal division between these two energies. The kinetic energy is due to the motion of the atoms and is dependent on temperature. The potential energy is the energy stored in the bonds of the atoms as they compress and extend, as in a spring, which coils and recoils when pressed and released. In gases (see Chapter 2), much of the internal energy is due to kinetic energy as there is little storage of energy between gas atoms.

The kinetic energy in a solid can be increased if heat is applied. The definition of heat is 'the energy which is transferred from a body of higher temperature to one of lower temperature, by conduction, convection or radiation' (Duncan 1994: 78). The schematic in Figure 4.1 shows what happens when a solid is heated.

The transference of heat makes the atoms of a solid oscillate over a larger distance and with a greater speed. The larger range of motion of the atoms causes a greater mean distance between atoms, resulting in the solid expanding. The temperature increase also results in an increased vibration of the molecules due to their faster speed of movement.

However, for the internal energy of a solid to increase it is not solely a matter of heat application. It is also possible for work done on the solid to increase its internal energy. Thus the internal energy of a solid can be increased either by increasing the temperature and/or by the application of work. Therefore, the term 'heating' should be used carefully, and it might be that the term 'internal energy' as applied to a solid, liquid or gas is more appropriate.

The specific heat capacity is 'the quantity of heat required to produce a unit rise of temperature in unit mass' (Duncan 1994: 79). The units of measurement for specific heat capacity are joules per kilogramme per kelvin. The mean specific heat capacity for aluminium is 910 J·kg^{-1}·K^{-1}, for lead 130 J·kg^{-1}·K^{-1} and for water at room temperature 4200 J·kg^{-1}·K^{-1}. In textbooks it is also common to read about the heat (or thermal) capacity of a solid. This is defined as 'the quantity of heat needed to produce one unit rise of temperature in a body' (Duncan 1994: 79). The term 'specific' preceding heat capacity denotes a quantity that is standardised as 'per unit mass'. The specific heat of the human body is very close to that of water owing to the body's very high water content (~60%).

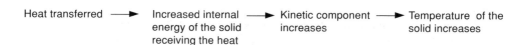

Heat transferred ⟶ Increased internal energy of the solid receiving the heat ⟶ Kinetic component increases ⟶ Temperature of the solid increases

Figure 4.1 Increased internal energy of a solid through heating

Transfer of heat in a solid

A further characteristic of a solid is the ability to transfer heat in the form of **conduction**. Conduction occurs when molecules of two surfaces contact one another. Conduction allows for the direct transfer of the motion of molecules in a hot area to molecules in a cooler area. Hence, where there is a thermal gradient heat will move from a hotter region to a cooler region. At the molecular level, the hotter molecules, which are moving more rapidly due to the increased temperature, are also resulting in more collisions. As a consequence, the rapidly moving hotter molecules give internal energy to the molecules of the cooler regions. Many people are aware that metals are better conductors of heat than non-metals but are not necessarily aware of the reason why. A simple experiment such as heating a piece of metal and a piece of wood to the same temperature (e.g. 70°C) will help to elucidate the process of conduction in solids. In a piece of heated metal some of the electrons are more free to move around, in contrast to the more rigidly fixed electrons in a piece of wood. In the metal, the electrons will be moving at great speeds and will transfer energy by colliding with other electrons in the lattice. When a hand touches the metal, heat is transferred to the hand and the metal feels hot. When a hand touches the wood, although it is heated to the same temperature as the metal, the wood does not feel as hot. The number of collisions and the amount of energy transferred during each collision are less for the heated piece of wood. The relationship between the number of collisions and the amount of energy transferred is known as the **thermal conductivity**. Metals have a higher thermal conductivity than non-metals because they are more efficient during their energy transference as a result of the number of collisions of atoms. The thermal conductivity of water is about 25 times greater than air; hence it provides a greater transfer of heat. If the simple example with a piece of metal and wood was changed so that the wood and metal were cooled to the same temperature (e.g. 5°C), what would be the conclusion? When the metal is touched it feels much colder than the wood, and for the same reason that the metal felt hotter than the wood – the metal conducts heat more rapidly away from the hand compared with the wood. This experiment highlights an interesting fact about the human perception of heat. That is, despite the fact that both the metal and the wood were heated to the same temperature, the human perception is that the metal felt both hotter and colder respectively.

The conduction of heat is dependent on a number of other factors, including the relative masses of the two bodies, the thermal resistance of the contacting surfaces, the thickness of the conductor, the temperature differences between the surfaces, and the area of contacting surfaces. The conduction of heat can be represented by an equation known as Fourier's law of heat flow (Equation 4.1), measured in $W \cdot m^{-2}$:

$$H_k = k/d \cdot (T_1 - T_2) \cdot A, \qquad (4.1)$$

where H_k is the specific heat of the substance, k is the relative mass of the two bodies and thermal resistance between the two contacting surfaces, d is the thickness of the conductor, $T_1 - T_2$ is the temperature difference between the two surfaces, and A is the area of contacting surfaces.

Another possibility for heat transfer by a solid is **radiation**. Whereas conduction requires contact to transfer the energy from one molecule to another, radiation does not, as the transfer is not due to the temperature of the medium. The energy that the earth receives from the sun is a clear example of radiation (i.e. solar radiation). Radiation is independent of air temperature and the presence of moisture. This is why it is still possible to become sun-tanned on a cool cloudy day in summer. The equation for calculating radiation in a human body is as follows:

$$R = \sigma\varepsilon(A_r/A_d)\cdot(T_{sk}^{\ 4} - T_r^{\ 4}),\tag{4.2}$$

where σ is the Stefan–Boltzmann constant (5.67×10^{-8} W·m^{-2}·K^{-4}), ε is the emissivity of the body surface (typically 0.99), A_r/A_d is the fraction of skin involved in the radioactive heat exchange (typically 0.696 for sitting and 0.725 for standing position), T_{sk} is the mean skin temperature in kelvins (K), and T_r is the mean environmental radiant temperature in kelvins (K).

The emissivity of an object depends on how effective that object is as a radiator. An ideal radiator is also an ideal absorber and has a score equal to one, whereas an ideal reflector has a score of zero. The skin is an ideal absorber and radiator, absorbing between 97 and 99% of the infrared radiation that strikes it. The smoothness and thickness of the surface influences emission and absorption of radiation. Differences in skin colour do have a bearing on the amount of reflection of the sun's radiation in the visible light range, but there is no difference in absorption of infrared heat, with skin of all colours absorbing about 97%. The amount of heat lost by radiation will be examined in the application section below as it has a major impact on the amount of heat lost while the human body is at rest.

The processes of conduction and radiation become relatively less important when exercise is taking place in hot environmental conditions. In such conditions, evaporation becomes the major route of heat loss.

All solids will radiate heat, but since this heat is in the infrared spectrum it cannot be seen by the naked eye. The technique that allows recording of infrared radiation is known as thermography. The recorded image taken by the thermal camera is known as a thermogram. The image distinguishes different temperature zones, known as isotherms, by showing them in different colours or shading. Special imaging cameras can be used to find people lost in open countryside where undergrowth might restrict visibility, or at night when visibility is poor. In physiology laboratories it is possible to use thermal imaging cameras to investigate the effects of heat on the body whilst exercising.

Categories of a solid

There are three types of solid:

- crystalline (which can be subdivided into four structures: faced centred cubic; hexagonal close packing; body centred cubic; tetrahedral);
- amorphous;
- polymers.

Almost all solids, including all metals and most minerals, are crystalline. The key features of crystals are the repetitive nature and regularity of the patterning of atoms, ions or molecules (see Figure 4.2). In crystals which are described as faced centred cubic (FCC), such as sodium chloride (table salt), the crystal is shaped like a cube. At the centre of each of the six faces of the cube is a particle. Therefore, each chloride ion is surrounded by a sodium ion and similarly every sodium by a chloride ion. In hexagonal close packing (HCP), as the term implies, the structure of the packing is hexagonal; zinc and magnesium are two examples of

Face centred cubic

Hexagonal close packing

Body centred packing

Tetrahedral

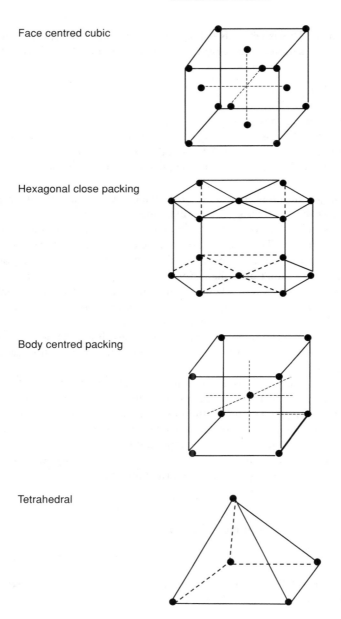

Figure 4.2 The four types of crystal

such structures. In body centred cubic (BCC) crystals, the structure is less tightly packed than that of FCC. Although similar to the cube structure of the FCC crystal, the BCC has only a single particle in the centre of the cube and one particle in each corner. Alkali metals are typically of this structure. The tetrahedral structure is a more open arrangement, and has a particle in the centre and one at each of the four corners of the tetrahedron. Structures like these include carbon in the form of diamonds, silicon and graphite.

Solids in the second class, amorphous structures, do not contain the same organised pattern as crystals. It is only for a small volume that there is any order to their pattern; over large volumes there is a lack of any recognisable regular patterning. Glass is a common amorphous solid.

Solids of the third type, polymers, characteristically have large molecules, ranging from 1000 to 100 000 atoms, and are usually organic (carbon) compounds. Polymers may exist naturally (e.g. cellulose found in plants) or be manufactured (e.g. Perspex, nylon or polyester). Natural polymers possess a repeating and regular patterned structure of so-called monomers. Cellulose, for example, comprises anything from several hundred to several thousand glucose molecules arranged in a long chain. The bicycle used by Chris Boardman to win the gold medal in the 4 km pursuit in the 1992 Barcelona Olympics was carbon fibre-reinforced plastic. This offered greater stiffness and strength than the plastic materials previously used, which were often reinforced with glass (otherwise known as fibre-glass). Although the carbon-reinforced plastic is much lighter than a conventional metal-framed bicycle, the material is very costly and therefore its use is not widespread for lower-level athletes. Tennis is another sport that has benefited from the use of reinforced plastic; it is very rare for a professional player to compete with a racket made of anything other than carbon fibre plastic. Carbon fibre plastics are extremely strong and resilient, yet are light enough to allow a player control of a powerfully struck tennis ball.

APPLICATION OF SCIENCE TO EXERCISE AND SPORT

1 Radiation from the body at rest

Regulation of the amount of heat gained or lost is important to the survival of the human body. The body maintains its temperature within a narrow band, approximately 36–37.5°C (97–99.5°F). There is, however, some variation due to individual differences, such as the effects of the menstrual cycle in women (typically body temperature is 0.5°C higher after ovulation) and time of day (body temperature is lowest in the early hours of the morning and highest in the early evening). Depending on the environmental conditions, the body must therefore regulate the amount of heat lost or gained (see Figure 4.3). The importance of this is indicated by the fact that the body need experience only a 6°C increase in temperature to make the risk of death from hyperthermia a very serious possibility.

The large majority of chemical processes occurring in the human body result in the release of heat in addition to useful work being done (see Chapter 6). At rest, the very metabolically active organs (i.e. the heart, kidney, liver and brain) produce the majority of the body's heat, whilst muscle and skin account for about 20%. The basal metabolic rate (**BMR**) represents the quantity of internal energy that must be used to maintain normal bodily functions during the waking state (i.e. biological maintenance of the body at rest). Typically, the average BMR for an adult male is 40 kcal·h^{-1}·m^{-2}. So for a male whose body surface area is 1.8 m^2, this would equate to 72 kcal·h^{-1}, or approximately 83 W. (Females generally have a lower BMR as on average they are smaller than males, but the figures are quite similar.) Hence, for an average person approximately 83 W of heat would be gained every hour; because heat balance is generally maintained, the same amount of heat must also be lost. If the body did not get rid of the heat, the metabolic processes occurring at rest would increase the body's temperature by about 1°C per hour. It would therefore take only a matter of hours before the

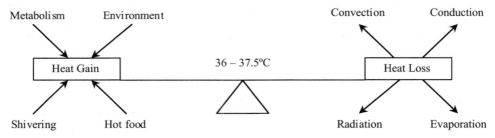

Figure 4.3 The heat balance of the human body through heat loss and heat gain

body would be in danger from excessive temperature. This would be particularly likely when the environmental temperature is greater than the body temperature.

As shown in Figure 4.3, one of the ways the human body can lose heat is through radiation. Radiation is a very effective means of body heat loss. To highlight how much radiant heat can be lost by a human, take the scenario of an unclothed body with a skin temperature of 32°C and an air temperature of 24°C (for simplicity we are assuming the temperature is uniform around the body). Using Equation 4.2:

$$R = \sigma\varepsilon(A_r/A_d)\cdot(T_{sk}^{\ 4} - T_r^{\ 4}),$$

where
σ is the Stefan–Boltzmann constant = 5.67×10^{-8} W·m^{-2}·K^{-4};
ε is the emissivity of the body surface = 0.99;
A_r/A_d is the fraction of skin involved in the radioactive heat exchange = 0.725 in standing position;
T_{sk} is mean skin temperature = 32°C*;
T_r is mean environmental radiant temperature = 22°C*.
(*Note that in the equation temperature is in kelvins.)
Thus:

$$R = (5.67 \times 10^{-8}) \times 0.99 \times 0.725 \times (305^4 - 295^4)$$

$$= (5.67 \times 10^{-8}) \times 0.99 \times 0.725 \times (1.08 \times 10^9)$$

$$= (5.67 \times 10^{-8}) \times 0.717 \times (1.08 \times 10^9)$$

$$= 61.25 \times 0.717$$

$$\approx 44 \text{ W·m}^{-2}.$$

If the unclothed body had a surface area of 2 m², then the loss of heat due to radiation would be approximately 88 W. Therefore, it can be seen that the heat gained by the body is effectively being lost again by radiation and so the body is essentially in thermal balance. If the temperature difference between skin and air were widened, then the heat lost by radiation would be greater than the heat gain of the body. In this situation, the body would begin to feel cold. The natural reaction by the body would be to vasoconstrict blood vessels near the surface of the skin so that heat would not be as readily lost. In addition, shivering could be initiated to increase the internal energy of the body. For further information on the effects of thermoregulation, and more calculations which include the addition of conductive, convective and evaporative losses, see Eston and Reilly (1996), chapter 12.

2 Bone as a solid

Bone is not often thought of as an active tissue within the body. However, it is in fact a dynamic tissue and throughout a human's life is constantly being modelled and remodelled. Bone can be subdivided into two structural types: cortical and cancellous (trabecular) bone. Cortical bone is the denser of the two and is found on the external surfaces of bones and in the walls of the diaphyses (bone shafts). The thickness of the cortical bone varies as a function of the mechanical requirements. Cortical bone is therefore strong and resistant to bending. Trabeculae bone is found within the vertebral bodies, in the epiphyses (growing parts) of long bones and in short bones. Irregular spacing found in trabeculae bone reduces the weight of the bone, and the orientation of the bone structure is often aligned with the forces impacting on the bone.

Bone is a composite of water (25–30%), protein in the form of collagen (25–30% of the bone's dry weight) and minerals, mostly calcium phosphate and carbonate (65–70% of the bone's dry weight). The role of the collagen is to resist tensile loading, that is, loading that stretches the bone along its longitudinal axis. The collagen fibres have limited flexibility, whilst the proteins act to give ductility (i.e. flexibility or ability to deform) and toughness in the form of resistance to shock loading. The minerals, calcium and phosphate, give resistance to compressive forces and result in hardness and rigidity of the bone. The rigidity is due to the minerals calcium phosphate and calcium hydroxide, which form calcium hydroxyapatite $(3Ca_3(PO_4)_2Ca(OH)_2)$.

When mechanical force is applied to a bone it will temporarily bend. This temporary change in shape can be measured as stress, which can be defined as the relative change in length (see Chapter 5). The amount of stress is dependent on several factors, such as the size and direction of the force, the distance between the applied force and the bending point, and the moment of inertia of the bone. Consider the compressive stresses of the tibia as a person walks or runs. The amount of force being absorbed by the body when running can be up to five times the person's body weight. This will be combined with shear stress that is caused by torsional loading associated with lateral and medial rotation of the tibia. The tensile stresses are much greater in running than in walking, but the shear stresses are greater in walking. In all sport and exercise situations there will be mechanical forces experienced by bone. In cricket, for example, as a batter goes to play a forward defensive shot, he is required to lunge forward quickly on to the front foot. The act of lunging will load the tibia and fibula. The bowler will also experience large impact forces which must be absorbed by the vertebral bones, and it is of no surprise that many bowlers suffer from bad backs throughout their careers. At present, much emphasis is being placed on videotaping bowlers to analyse their bowling technique in an effort to correct poor bowling actions and hopefully prevent future back injuries.

Mechanical stress on bone stimulates both cellular and tissue reactions by promoting the activity of hormones such as prostacyclin and prostaglandin E_2 and enzymes such as glucose-6-phosphate dehydrogenase. Without mechanical stress placed on bone, as is the situation for astronauts in space who experience weightlessness, bone mass is lost and optimal conditions for bone modelling and remodelling are reduced. By contrast, studies on athletes such as tennis and baseball players have found that the bones in their racket-holding and throwing arms have a higher bone density than those in the non-dominant arm. Swimmers have also been shown to have different bone densities, exhibiting a lower vertebral bone mineral density when compared with other athletes and controls (Taaffe *et al.* 1995). This is one of the reasons why swimming is not promoted as an exercise for optimal bone health. It is still unclear exactly how mechanical stress alters bone density, but the dominant theory at present compares bone to a piezoelectric crystal. The piezoelectric crystal produces a voltage when it is subjected to forces which deform the crystal. The voltage is then thought to stimulate the osteoclasts that lead to calcium synthesis.

A fracture is one of the most traumatic events to happen to bone. In sport and exercise the chances of this occurring are higher if the sport involves physical contact (e.g. football and rugby). Fractures tend to occur as a result of acute excessive stress, whilst stress fractures are the result of chronic excessive stress. Any force acting on a bone in excess of the force limits of the bone can result in a fracture. It is thought that shin splints are often caused by excessive amounts of landing on hard surfaces (e.g. running) and are due to micro-fractures of the tibia bone pulling the muscle tissue away from the periosteum of the bone. New recruits in the armed forces have been found to suffer from what is termed 'fatigue fractures'. These are the result of the sudden increases in physical training, which place chronic forces on the bone.

Fractures can more easily occur if the bone is weak. This is particularly evident in people with a condition called osteoporosis, also described as 'porous bone', 'brittle bones' or 'too little bone in the bones'. Osteoporosis typically results in low bone mass and a weakened architecture of the bone. The underlying weakened structure of the bone leads to increased risk of fracture. Fractures tend to be most prominent in the vertebrae, distal radius (forearm) and hip bones. The disease affects both males and females but is far more common in the latter. At present the role of physical activity is being promoted as a method to aid bone health, particularly in young teenage girls. This strategy is an attempt to optimise the bone mineral density, which is thought to reach a peak around the age of 30. For a comprehensive review of osteoporosis, and the impact of physical activity, see Drinkwater *et al.* (1995). The increasing number of people suffering from osteoporosis has resulted in the American College of Sports Medicine (1995) issuing a position statement about the disease. Four key points are highlighted:

1   Weight-bearing physical activity is essential for normal development and maintenance of a healthy skeleton.

2   Sedentary people may increase bone mass by becoming more active. However, the benefit of becoming active may be one of avoiding further losses of bone through being inactive.

3   Exercise is not a substitute for hormone replacement therapy at the time of the menopause.

4   An optimal programme for the elderly would include strength, flexibility and coordination exercises.

ACTION POINTS

1   Go through Equation 4.1 and input a variety of different skin and air temperatures to calculate the effect on radiation loss.

2 Why are runners sometimes given a silver foil blanket after completing an event like a marathon?

3 Why are runners still hot even an hour after finishing exercising?

4 What causes rickets and how does it affect bone?

5 Put the following in order of the highest mechanical loading for the promotion of bone health: cycling, running, swimming, weight training.

Conclusion

The factors that impact on a solid are simpler than those of a liquid and a solid's behaviour is more predictable. In a solid the atoms vibrate around a fixed point and successively repel and attract neighbouring atoms. The alternate movement of atoms reflects the internal energy and is a consequence of the potential and kinetic energy available within the solid. The fixed position of the atoms also gives the solid its characteristic shape and rigidity.

There are three main types of solid: crystalline, amorphous or glassy, and polymers. Subjects such as mechanical engineering investigate the external forces imposed on the solid and its subsequent behaviour. Mechanical properties such as strength, stiffness, ductility and toughness are important factors to consider in the construction of materials. In recent years information from an engineering perspective has been incorporated into sport and exercise science. In particular, biomechanists have routinely argued for an engineering approach or model within which to investigate many problems faced in sport and exercise.

An important aspect of solids is their ability to transfer heat. This is achieved in one of three ways: radiation, convection or conduction. At rest, radiation would appear to be the most important mechanism by which heat is lost from the body. During exercise, however, radiation, convection or conduction may not be adequate heat loss mechanisms, and in this instance evaporation would be the most effective method.

Bone demonstrates many of the properties of solids, i.e. strength, stiffness and toughness. It is important, however, not to neglect the dynamic nature of bone as a living tissue. Habitual physical activity subjects bones to stresses and strains, which promote bone growth and remodelling. In some diseases, such as osteoporosis, the mechanical property of toughness is severely compromised. One strategy that can be used to reduce the number of osteoporosis cases is to promote physical activity throughout life, with a particular emphasis on adolescent and menopausal women.

KEY POINTS

- Most solids have the appearance of rigidity and do not take the shape of containers in which they are placed.
- Since the particles within the solids are held in approximately fixed positions, there is less empty space in which atoms can move.
- The atoms of a solid vibrate to and fro around a fixed point. This vibration and the potential energy of the bonds of the atoms is known as internal energy.
- At room temperature, the average kinetic energy as indicated by the speed of the molecules of a solid is equal to several hundred miles per hour.
- Heating is the transfer of energy from a body of higher temperature to one of lower temperature by radiation, convection or conduction.

- The specific heat capacity of a solid is the quantity of heat required to increase one unit of mass by one unit of temperature.
- Heating a solid increases the vibration of the molecules and the solid expands.
- The greater the number of collisions of atoms, and therefore the amount of energy transferred by a solid, the higher the thermal conductivity.
- Conduction requires contact to be made between the molecules of two surfaces in order for heat to be transferred.
- Factors such as the mass and thermal resistance of the contacting surfaces, the thickness of the conductor, the temperature gradient between the two surfaces, and the area of connecting surfaces must also be considered when discussing conduction.
- Silver and copper are considered to be the best conductors, while polythene is the worst.
- Transfer of heat energy by radiation does not require contact and does not involve the medium that it travels through. An example is the radiation from the sun.
- At rest, radiation accounts for a large percentage of the heat loss from the body.
- There are three kinds of solid: crystalline (faced centred cubic, hexagonal close packing, body centred cubic and tetrahedral), amorphous and polymers.
- The human body is homeothermic, that is, it must maintain a thermal balance. If thermal balance is not maintained, death can occur.
- The use of different solids in sports equipment has had a major impact on the development of these sports.

Bibliography

American College of Sports Medicine (1995) Position stand on osteoporosis and exercise. *Medicine and Science in Sport and Exercise* 27(4): i–vii.

Drinkwater, B.L. *et al.* (1995) C.H. McCloy Research Lecture: Does physical activity play a role in preventing osteoporosis? *Research Quarterly in Exercise and Sport* 65: 197.

Duncan, T. (1994) *Advanced Physics*, 4th edition. London: John Murray Publishers.

Eston, R. and Reilly, T. (1996) *Kinanthropmetry and Exercise Physiology Laboratory Manual. Tests, Procedures and Data*. London: E & FN Spon.

Taaffe, D.R., Snow-Harter, C., Connolly, D.A., Robinson, T.L., Brown, M.D. and Marcus, R. (1995) Differential effects of swim versus weight bearing activity on bone mineral status of eumenorrheic athletes. *Journal of Bone Mineral Research* 10: 58.

Uvarov, E.B. and Isaacs, A. (1986) *The Penguin Dictionary of Science*, 6th edition. London: Penguin Group.

Further reading

Pascoe, D.D., Shanley, L.A. and Smith, E.W. (1994) Clothing and exercise. I: Biophysics of heat transfer between the individual, clothing and environment. *Sports Medicine* 18(1): 38–54.

Part 2

FORCE, PRESSURE, ENERGY
AND ELECTRICITY

5

FORCE AND PRESSURE

AIMS OF THE CHAPTER

This chapter aims to provide an understanding of the scientific principles of force and pressure which are relevant to the theory which underpins sport and exercise. After reading this chapter you should be able to:

- define the units of force and pressure;
- explain the importance of force and pressure in exercise and sport;
- apply the principles of force and pressure to exercise and sport.

Introduction

In the study of exercise and sport, it is the effect of force application on the human body which is of interest. In engineering, the usual safety factor built into materials and associated physical structures is over two times the force that is applied on a regular basis. This level of caution is known to result in a low probability of a failure caused by mechanical stress. Similar safety factors are built into materials and structures that have to withstand pressure (i.e. force per unit area), such as gas cylinders. An extreme engineering example is a lift, where it is particularly important that failure does not occur! A cable supporting the lift can usually withstand 11 times the maximum load specified by the manufacturers (Diamond 1993). Interestingly, the human body also seems to have been designed (evolved) with a safety factor of at least two in many of its structural and physiological systems. Basic examples include the lungs and the kidneys, of which we have two when one would be enough. In relation to forces, bones and tendons do not break until forces of two to five times peak natural force have been applied (Diamond 1993).

Throughout this chapter the emphasis will be on the effect of the application of force and pressure on the human body. Once the concepts of force and pressure have been defined, the various units used in measurement of force and pressure will be examined. The application of force and pressure provides a mechanical stress, and the consequences of such stress will be addressed. Since forces can only be applied to solid objects, whereas pressure can be applied to all physical states, the information in this chapter is closely related to that in the three chapters forming Part 1.

Scientific principles of force and pressure

Movement of an object (i.e. motion) depends on the application of **force**. Another way of stating this is to say that a change in motion is the visible evidence of the application of force. Sir Isaac Newton (1642–1727) defined the laws of motion, and hence the laws that govern the effect of forces. These laws are important in the study of exercise and sport, and relate to some of the information in this chapter. They are given in Bartlett (1997), and can be summarised as follows:

- Newton's first law, the law of inertia: a force is required to stop, start or alter the motion of an object.
- Newton's second law, the law of acceleration: the acceleration of an object is proportional to the force acting on it, and takes place in the direction of the force.
- Newton's third law, the law of action–reaction: to every action created by a force there is an equal and opposite reaction.

When applying these laws it should be remembered that they are true only for **rigid bodies**. Since the human body is not a rigid object, the laws should be applied with caution.

Physical states, force and pressure

Forces can be applied to solids. For rigid solids, forces can be applied in the form of compression or tension forces. **Compression** forces act to decrease the length of a solid, whilst **tension** forces act to increase the length (i.e. stretch a solid). The effect of a force (i.e. stress) acting on a solid is known as **strain**. The degree to which a solid changes length as a result of an applied force gives the strain. Tension forces, but not compression forces, can also be applied to non-rigid solids, such as a piece of elastic (see Chapter 4).

Pressure can act on a solid, liquid or gas. However, gas must be contained before pressure can be exerted. The **ideal gas** laws (see Chapter 2) illustrate the effect of pressure on a gas.

Units of force and pressure

A quantity with a magnitude (i.e. size) only is known as a **scalar** quantity. Force is a **vector** quantity, which means it has a directional component in addition to a magnitude. A force is visualised through the action it has on an object. Specifically, **acceleration** is the kinetic (i.e. movement) effect associated with the action of a force. The magnitude of a force is equal to the product of mass (kg) and acceleration ($m \cdot s^{-2}$) (see Equation 5.1). The unit of measurement of force is the **newton** (N).

$$\text{Force (N)} = \text{Mass (kg)} \times \text{Acceleration } (m \cdot s^{-2}). \tag{5.1}$$

Whilst forces are important in exercise and sport, it is often a combination of the magnitude of the force and the time over which it acts which determines performance. The product of force (N) and time (s) is known as **impulse** (Ns):

$$\text{Impulse (Ns)} = \text{Force (N)} \times \text{Time (s)}. \tag{5.2}$$

Pressure is the force applied per unit area, and is calculated by dividing force (N) by area (m^2) (see Equation 5.3). The unit of measurement of pressure is the **pascal** (Pa), where 1 Pa is equal to $1 \ N \cdot m^{-2}$.

Pressure (Pa) = Force (N)/Area (m^2). (5.3)

Although the pascal is the unit of pressure that is recognised by the international system of units, other units of pressure are widely used. For example, pressure is often measured with a column of liquid, where the height of the liquid in the column reflects the pressure exerted upon it. Since mercury is a very heavy liquid, it is often the liquid of choice. This is because with a heavy liquid a smaller column is required to demonstrate graduations of pressure. Consequently, pressure is often recorded in millimetres (mm) of mercury (Hg) (i.e. mmHg). Since atmospheric pressure is often determined with a column of mercury, the usual units used are mmHg. It is known that atmospheric pressure varies around 760 mmHg at sea level. The unit barometric pressure (bar) is related to the unit mmHg, with 1.01325 bar being equivalent to 760 mmHg (see Equation 5.4). The bar is frequently used when measuring ambient pressure in various environments, especially high-pressure environments such as occur in diving. There are 760 mmHg to 100 000 pascals (or 100 kPa) (see Equation 5.5).

760 mmHg = 1.01325 bar. (5.4)

760 mmHg = 100 000 Pa. (5.5)

For further information about units, particularly units of pressure, consult appendices in the back of standard physiology textbooks such as Astrand and Rodahl (1986), McArdle *et al.* (1996), Powers and Howley (1997), and Wilmore and Costill (1999).

Force

In addition to accelerating an object, the application of force is known to place **mechanical stress** on solid objects. The stress (Pa or N·m^{-2}) can be applied as either a **compression** force or an **extension** (tension) force, and is calculated by dividing force (N) by cross-sectional area (m^2) (see Equation 5.6).

Stress (Pa) = Force (N)/Area (m^2). (5.6)

The way in which an object responds to the application of force is linked to the elasticity of the structural material. The degree of **elasticity** is known as the **modulus of elasticity** (given the symbol E), and is equal to the applied stress (given the symbol σ) divided by the resulting strain (given the symbol ε) (see Equation 5.7). The applied stress is measured in units of pressure (Pa) and the resulting strain is measured as the percentage (%) change in length (see Equation 5.8):

Elasticity (E) = Stress (σ)/Strain (ε), (5.7)

where

Strain (%) = Change in length (m)/Original length (m). (5.8)

In the case of tension forces, a solid object usually has a limit of tolerance, known as the **tensile limit**, above which the material fractures. Prior to this point, the material will pass from an elastic region of deformation to a **plastic** region of deformation. The point of transition from plastic to elastic deformation is known as the **yield point**. In the elastic region the structure of the material is deformed by the tension force and then returns to its original structure when the force is removed. In the plastic region of deformation the structure of the material is deformed irreversibly, resulting in a permanent length change of the object.

Some solid materials can be placed under tension but not compressed. Elastic is a good example. Within the human body tendons, ligaments and muscle can all be placed under tension, but not compressed.

The human body has many properties of a solid in that the various parts can be placed under tension *and* compressed. However, as is the case with many solids, the human body is not rigid. Mechanical stress placed on the human body cannot, therefore, be treated as though forces are being applied to a rigid body. The consequence, as mentioned earlier, is that Newton's laws of motion cannot always be applied.

In the case of the human body, as with other solid objects, it is functionally useful to consider whether forces applied are repetitive or isolated events. The technical phrases are **chronic** mechanical stress for repetitive force application, and **acute** mechanical stress for isolated force application. Thus a single impact, such as an unintentional collision, would be referred to as acute stress, whilst repeated impacts, such as those experienced when running, would be referred to as chronic stress. Although the process of application of the force may differ, the end result of exceeding the tolerance of a particular material is usually similar. In the human body, for example, the consequence of exceeding the tolerance of a part of the body is an injury. An injury which develops over the short term, such as one occurring during a collision, is known as an acute injury, whilst an injury that develops over a longer period is known as a chronic injury.

Types of force

When studying the human body during exercise and sport it is often more appropriate to consider **torque** (or turning moment) rather than force. Torque is the product of force (N) acting on an object and the perpendicular distance (m) between the point of application of the force and the centre of rotation of the object (see Equation 5.9). Torque is often a more appropriate measure since the human body consists of many joints or **levers**. Force produced by the muscles is transferred to external objects through these lever systems.

Torque (Nm) = Force (N) × Perpendicular distance (m).   (5.9)

Various other types of force are known to act on the body during exercise and sport. **Drag** is the force opposing the motion of a body travelling through a liquid or gas. When an individual is moving through air, during running for example, a drag force due to the still air is experienced. This drag force is sometimes increased due to the presence of a head wind, the velocity of which should be added to the velocity of the runner when calculating drag force. Depending on the type of fluid (i.e. liquid or gas), various components of a drag force are important. Pressure drag is the opposing force due to the negative pressure created behind a moving object. Wave drag is the force opposing motion due to the creation of a wave on the surface of a liquid. Surface drag is the force opposing motion due to the tension on the surface of a liquid.

Frictional forces are an important consideration during exercise. For example, the only way in which an individual can translate the force produced by muscle contraction into effective locomotion is with the aid of frictional and gravitational forces. Frictional forces are markedly reduced when walking or running on ice! A frictional force increases in response to an opposing force (i.e. Newton's third law: every action has an equal and opposite reaction). Eventually the opposing force is increased to such an extent that the object starts to slide. As sliding takes place the frictional force is reduced, but still presents an opposition to motion. The frictional force during sliding is known as sliding friction.

Resolving forces

When attempting to quantify the magnitude and direction of a force (i.e. a force vector), two approaches can be taken. The forces can either be added together to give an overall force vector, a process known as composition of forces, or resolved in defined directions, known as **resolution of forces**. A graphical technique, which produces what is known as a force parallelogram, can be used to quantify a force vector. The parallelogram law states:

- If two forces acting at a point are represented in magnitude and direction by the sides of a parallelogram drawn from the point, their resultant is represented by the diagonal of the parallelogram drawn from the point.

Thus in this technique, forces are drawn from a point, with the magnitude of each force being proportional to the length of a line, and the direction of each force being indicated. By then creating a completed parallelogram, the diagonal line from the point of origin to the far corner represents the composed force vector (see Figure 5.1).

Forces can also be composed and resolved using mathematical techniques. By knowing the magnitude of the force and the angle between the direction of resolution and the direction of the force, the magnitude of a force acting in a particular direction can be quantified. The mathematics used to perform this transformation is known as **trigonometry**.

Measurement of force

Strain gauges are used to measure the magnitude of a force. (In the exercise and sport situation, however, as already mentioned, it is normally torque rather than force that is

Composition of Forces 1 and 2

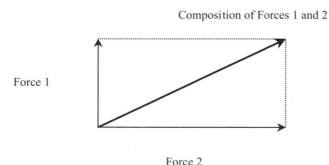

Force 1

Force 2

Figure 5.1 Composition of forces with the parallelogram technique

measured.) Various types of strain gauges are available. Force may be determined through electrical resistance, mechanical or optical principles. It is common to find an electrical resistance strain gauge in an exercise and sport laboratory. The basic principle of operation is as follows. By applying a force to a length of wire, the wire changes length and cross-sectional area changes accordingly. Since resistance is dependent upon length and cross-sectional area, these mechanical changes have an impact on the electrical resistance of the wire, which in turn affects the voltage output. A process of **calibration** is then used to relate particular voltages with known forces. See Application 1 below for an example of the use of such a strain gauge.

Pressure

Mechanical load can be applied through pressure. Although forces can apply a direct compressive load only on a rigid solid material, pressure can exert a compressive load on solids (whether rigid or not), liquids and gases. The similar effects of pressure on liquids and gases have been covered in Chapters 2 and 3.

Measurement of pressure

As mentioned earlier in the chapter, pressure is often measured with a column of liquid. Although mercury is often the substance of choice (because of its high density), other substances are also used. For example, when examining the pressure that can be produced during inspiration or expiration, a column of water is often used (see Figure 5.2).

The water naturally tends to find the point of least potential energy, and therefore levels out in the U-shaped tube, so that two columns of water of the same height reside either side of the U bend. If an individual then inspires or expires into one end of the tube, the inspiratory/ expiratory pressure can be determined from the change in height of the water (i.e. mmH$_2$O). Pressure in pascals is calculated by dividing the force (i.e. the mass of water moved against the acceleration due to gravity) by the cross-sectional area of the tube. The mass of water

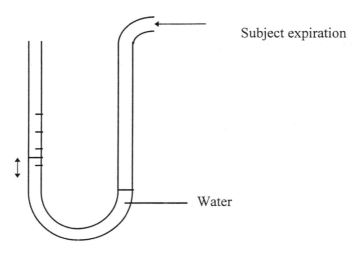

Subject expiration

Water

Figure 5.2 A manometer

moved is equal to the product of volume (i.e. height change × cross-sectional area) and density of the water (i.e. ~1 g·cm^{-3}).

APPLICATION OF SCIENCE TO EXERCISE AND SPORT

1 The relationship between joint torque and joint angle

The measurement of force production by the knee extensor muscles is a regular laboratory practical in the study of sport and exercise. Since it is the measurement of force about a fixed point (i.e. the knee joint) which is considered in such a practical, the correct term is torque. To determine torque, the force is measured using a strain gauge. The strain gauge is usually situated in a cable attached to both the subject's ankle and a support (see Figure 5.3).

The maximal torque produced by the knee extensor muscles is determined when the participant performs a maximal voluntary contraction (MVC). Maximal torque is often determined at a range of knee angles, and consequently a range of muscle lengths. In this situation, the torque is determined with the muscles working isometrically. **Isometric** muscle activity is the term used when the muscles are attempting to shorten, but fail to do so due to a high external load. Figure 5.4 shows typical results from a participant performing a series of maximal voluntary isometric contractions ranging from a flexed (30°) to an extended (150°) knee joint. The results show that the highest torque is produced when the knee is about halfway between flexion and extension (i.e. 90°). As the joint moves towards the extremes, the maximal isometric torque is reduced. Therefore, when maximal torque is discussed it is essential to make reference to the joint angle.

2 The relationship between joint torque and joint angular velocity

When examining the torque produced by a joint, it is necessary to consider the velocity of shortening of the muscles in addition to the muscle length. Since torque about a joint is

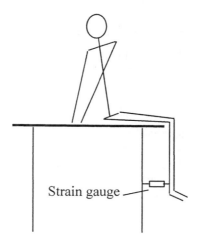

Figure 5.3 Measurement of torque about the knee joint

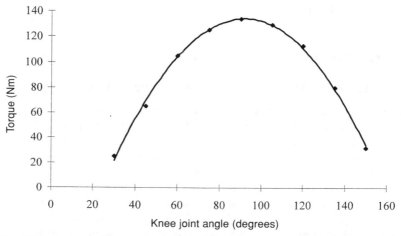

Figure 5.4 The relationship between torque and knee joint angle during maximal voluntary isometric contractions of the knee extensor muscles

being studied, it is practical to consider the angular velocity of the joint rather than muscle shortening velocity. In Application 1, an angular velocity of $0°·s^{-1}$ was considered because the contraction was isometric (i.e. there was no shortening of the muscles). Equipment is now available to determine force or torque production when a joint angle is changing with a pre-determined velocity (e.g. Cybex ergometer). The result of a series of trials at differing velocities is a torque–velocity relationship. A typical such relationship is shown in Figure 5.5. Torque is normally highest at the lowest velocity, and decreases in a curvilinear fashion as velocity increases. In fact, the highest torque is produced when a muscle is lengthened, but resisting the lengthening. A muscle contraction at a pre-determined velocity is known as an **isokinetic** contraction. Normal muscle activity when the velocity of contraction is constantly changing is known as **isotonic** contraction.

3 Assessment of the ground reaction force

A force plate may be used to determine the ground reaction force to an object moving over the ground. In the study of exercise and sport, a force plate is often used to examine the ground reaction force due to a participant walking, running, or performing jumping movements.

A **force plate** is essentially a metal plate inserted into the floor which lies flush with the floor surface. The plate is supported by four pillars, one on each corner, with each pillar housing a strain gauge, known as a foil or wire gauge. Since the electrical resistance is proportional to the length of the gauge, the electrical resistance provides a measure of the force (see Figure 5.6). Each pillar can measure force in three directions (i.e. vertically and in two horizontal directions). The vertical force is usually labelled as F_z, the horizontal force in the direction parallel to the motion as F_y, and the horizontal force perpendicular to the motion as F_x (see Figure 5.7).

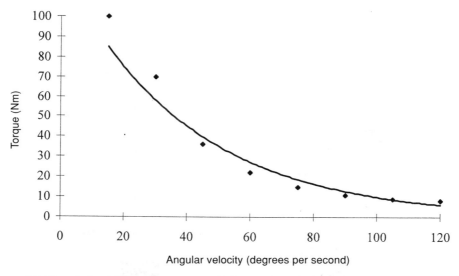

Figure 5.5 The relationship between torque and joint angular velocity during maximal voluntary knee extension exercise

Typical traces for a participant running over a force plate are shown in Figure 5.8. Graph (a) in the figure shows F_z in a single stance phase during running. The first peak is usually observed as the foot strikes the ground, followed by the second peak as the muscles in the leg work against the acceleration due to gravity which is pulling the runner towards the ground. In graph (b) (i.e. F_y), the negative peak is the result of the initial deceleration (i.e. retarding force) following foot contact, and the positive peak is the result of final acceleration (i.e. propulsive force) prior to take-off. In graph (c) (i.e. F_x), the positive portion of the curve is the result of movement of the centre of gravity of the body over the foot in contact with the ground. Force traces such as those in Figure 5.8 have many clinical uses in addition to those in sport and exercise.

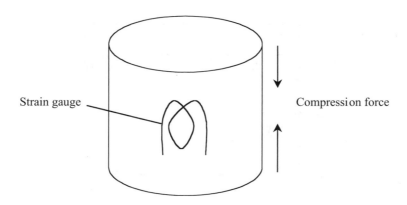

Figure 5.6 A strain gauge located within a pillar of a force plate

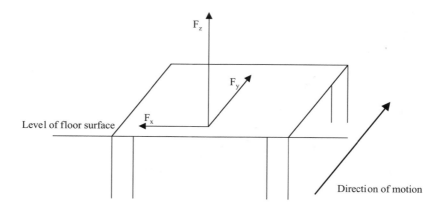

Figure 5.7 Force measurement in a vertical and two horizontal planes with a force plate

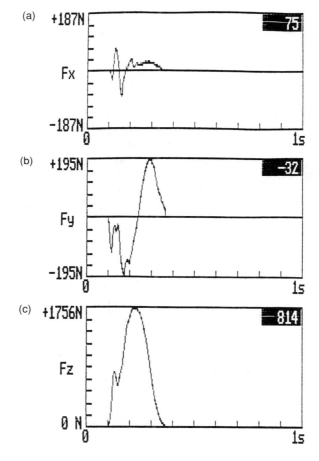

Figure 5.8 Typical force traces recorded in vertical (c) and horizontal (a and b) planes with a force plate

ACTION POINTS

1 Give the three laws of motion.
2 What visible signs show that a force is acting on an object?
3 What is the unit of measurement of force?
4 Which physical state can force act upon?
5 What variable is derived from force and time?
6 Give an example of a compression force on the body.
7 What type of solid can a compression force act upon?
8 Give an example of a tension force in the human body.
9 What is another term for a tension force?
10 How is force normally measured?
11 Give an example of limiting friction.
12 What is the difference between air resistance and wind resistance?
13 What is the modulus of elasticity?
14 How is change in length related to strain energy?
15 Under what conditions can pressure act upon a fluid?
16 Give three units for the measurement of pressure.
17 What is the relationship between force and pressure?
18 Draw and label a manometer and give one use of it.
19 What is a barometer?
20 Draw a graph of the relationship between force and muscle length.

Conclusion

The effect of force is observed through changes in motion, and hence Newton's laws of motion are important in understanding such an effect. However, Newton's laws are valid only for rigid bodies, so application of them to the human body should be done with caution. Force is the product of mass and acceleration of an object. When forces are generated by muscle contraction in the human body, the magnitude of these forces is normally measured as torque about a joint. Functionally, torque is important, since it is the product of force applied and distance between the joint and the point of application of the force. Force can be applied to solid objects, but compression force can be applied only to rigid solids, whereas tension force can be applied to all solids. Pressure (i.e. force per unit area) can be applied to solids and fluids (i.e. liquids or gases). With regard to sport and exercise, the time over which force is applied is important for the resulting motion; the product of force and time is known as impulse. In a sport and exercise laboratory, force between an object and the ground may be measured using a force plate. Additionally, muscle force production can be assessed through torque about a joint. With modern equipment, torque can be assessed when the muscle is working isometrically, isokinetically or isotonically.

KEY POINTS

- The effect of force is observed through the change in motion of an object.
- Force is equal to the product of the mass and the acceleration of an object.
- Newton's first law of motion is the law of inertia: a force is required to stop, start or alter the motion of an object.

- Newton's second law of motion is the law of acceleration: the acceleration of an object is proportional to the force acting on it, and takes place in the direction of the force.
- Newton's third law of motion is the law of action–reaction: to every action created by a force there is an equal and opposite reaction.
- The duration over which a force acts is functionally important, and the product of force and time is known as impulse.
- Force can act on a rigid solid as compression or tension, and act on a non-rigid solid as tension only.
- Pressure is force per unit cross-sectional area.
- Pressure can act on all states of matter, but on liquids and gases only if they are contained.
- Force is measured through strain gauges.
- Torque is the force operating about a point of rotation, and is calculated as the product of the force and the distance between the point of rotation and the point of force application.
- Torque is functionally important in sport and exercise since force is applied in the human body through a series of levers.
- Joint torque–velocity relationships give a functional insight into muscle force–shortening velocity relationships.

Bibliography

Astrand, P-O. and Rodahl, K. (1986) *Textbook of Work Physiology: Physiological Bases of Exercise*. Singapore: McGraw-Hill Book Co.

Bartlett, R. (1997) *Introduction to Sports Biomechanics*. London: E & FN Spon.

Diamond, J. (1993) Evolutionary physiology. In *The Logic of Life: The Challenge of Integrative Physiology*, C.A.R. Boyd and D. Noble (eds). Oxford: Oxford University Press.

McArdle, W.D., Katch, F.I. and Katch, V.L. (1996) *Exercise Physiology: Energy, Nutrition and Human Performance*, 4th edition. Baltimore: Williams and Wilkins.

Powers, S.K. and Howley, E.T. (1997) *Exercise Physiology: Theory and Application to Fitness and Performance*, 3rd edition. Boston: McGraw-Hill.

Wilmore, J.H. and Costill, D.L. (1999) *Physiology of Sport and Exercise*, 2nd edition. Champaign, IL: Human Kinetics.

Further reading

Dyson, G. (1985) *Dyson's Mechanics of Athletics*, 8th edition. London: Hodder and Stoughton (chapters 3 and 4).

Enoka, R.M. (1988) *Neuromuscular Basis of Kinesiology*, 2nd edition. Leeds: Human Kinetics.

Hopper, B.J. (1973) *The Mechanics of Human Movement*. London: Granada Publishing Ltd (chapters 3 and 4).

Kreighbaum, E. and Barthels, K.M. (1996) *Biomechanics: A Qualitative Approach for Studying Human Movement*, 4th edition. London: Allyn and Bacon (chapter 2).

6

ENERGY

AIMS OF THE CHAPTER

This chapter aims to provide an understanding of the scientific principles of energy, work and power which are relevant to the theory which underpins sport and exercise. After reading this chapter you should be able to:

- name the units of energy, work and power;
- describe the different forms of energy;
- explain the concepts of work, power and efficiency;
- apply energy-related concepts to exercise and sport.

Introduction

Energy is a very important consideration, not only in exercise and sport but also in daily activities. The human body relies upon a supply of energy to continue to move and to perform essential daily functions. Energy is needed for growth, repair, digestion and storage of food, and of course for physical activity. The energy required is supplied to the human body through food, hence the importance of good nutrition for optimum physical performance. Unlike plants, the human body cannot obtain energy directly from the sun (i.e. photosynthesis), and therefore relies upon plants, and animals that eat those plants, for energy. When considering energy, it is important to understand the implications of the first law of **thermodynamics**, which states:

Energy can neither be created nor destroyed, just changed from one form to another.

Throughout this chapter many examples will be provided of energy being converted from one form to another. When this process takes place in the human body, a significant amount of energy is usually converted to heat in addition to other forms of useful energy. The continual production of heat is one way in which the human body maintains a temperature of about 37°C, despite a usually cooler surrounding environment.

In this chapter there will be little consideration of how the body obtains its energy; this important area in the study of sport and exercise is covered elsewhere in detail (e.g. Brooks *et al.* 1995; McArdle *et al.* 1996; Maughan *et al.* 1997). Instead, emphasis will be placed on the different classifications of energy, and their potential in explaining physical performance. Further discussion of this topic may be found in Newsholme *et al.* (1994).

Scientific principles of energy

Units of energy

In the study of exercise and sport it is important to be able to quantify energy. Comparisons may then be made between how much energy exists in different forms, or how much energy is used by the body to do a certain amount of work.

The international system for classifying units expresses energy as a **joule** (J). One joule is the amount of energy required to displace the point of application of a force of 1 newton (N) through a distance of 1 metre (m) in the direction of the force (see Equation 6.1).

$$\text{Energy (J)} = \text{Force (N)} \times \text{Displacement (m)}. \tag{6.1}$$

However, the international system of units has not yet been fully adopted by every country in the world, so some countries (and individuals) still use the old unit of energy, the **kilocalorie** (kcal). A kilocalorie is the amount of energy required to heat 1 kg (i.e. 1 litre) of water by 1°C under standard conditions (i.e. from 14.5 to 15.5°C) (see Equation 6.2).

$$\text{Energy (kcal)} = \text{One litre water temperature change (°C)}. \tag{6.2}$$

Although it is not an ideal situation to have people using different units for expressing energy, the conversion between the two units is quite easy. There are ~4186 J (or 4.186 kJ) to a kcal, so a figure in kcal should be multiplied by 4186 to convert it into joules (see Equation 6.3).

$$1 \text{ kcal} = 4186 \text{ J}. \tag{6.3}$$

So, for example, if an energy drink contained 420 kcal per litre, it could be calculated that 1 758 120 J (i.e. 420×4186) or ~1758 kJ of energy was contained per litre of the drink.

Whilst the continued use of two different units for the measurement of energy is undesirable, this situation may unwittingly serve a useful purpose. This purpose relates to the definition of energy. The definition of energy often used – one related to the unit of energy of the international system of units, the joule – is 'the capacity to do **work**'. However, when applied to physical activity, this definition has limitations, since energy may be expended when no external work is performed (Winter 1990). Consider the situation when an individual is trying to lift an object that is too heavy. Although energy is expended, no work is done, since the object is not moved. So for the physical activity situation, a more appropriate definition of energy may be 'the capability to perform exercise' (Winter 1990). The fact that a unit of measurement of energy is still used (i.e. the kcal) that does not involve displacement in the line of action of a force in its definition is potentially useful in this regard.

Chemical energy

All chemical energy ultimately derives from the energy of the sun. Common forms of chemical energy that are used to supply energy to the home include coal, gas and oil, which are either used to generate electricity by first being converted to mechanical energy, or used to supply heat.

The idea that chemical energy is stored in food, and that the energy stored within the human body is in the form of chemical energy, is often less well understood. Energy in the form of food is digested by the body until it is in a form that the body can either use immediately, or transport to storage sites. Since the body does not have a continuous supply of energy from outside, unlike the energy supply to a home (e.g. natural gas for heat and cooking), the body must store energy for use at a later point. Once this energy arrives at the storage sites (e.g. glucose in the blood is delivered to the muscles), it is further transformed into the required form for storage (e.g. muscle glycogen). The stored energy within the body can then be released as and when required, to allow the body to perform mechanical work, and other processes such as the digestion of food, maintenance of a stable internal environment, repair, and growth.

The human body can directly utilise only one form of chemical energy. This is the energy that is contained within the bonds between phosphate and a substance known as adenosine (adenine and D-ribose). Adenine is usually bound to three phosphates via ribose and three high-energy bonds. This substance is known as a nucleotide, and is called **adenosine triphosphate** (ATP) (see Figure 6.1). As each of the high-energy bonds is broken, the energy released can be used by the body in a useful way. The fuel stores within the body are then used to re-form the broken high-energy bonds.

Since the human body has only a small supply of ATP, it must continually re-synthesise this useful form of chemical energy. In fact, if no re-synthesis took place the body could continue to exercise for only about 2 seconds before the ATP stores would be completely used up. Fortunately, the body is very adept at using the fuel stores to continually re-synthesise the ATP store.

To compare the amount of energy stored as ATP within a normal 70 kg body with other types of fuel, it is necessary to convert the size of all the energy stores within the body to the common unit of energy, the joule. Table 6.1 shows the amount of energy which is typically contained within each fuel store.

Clearly, the size of each fuel store within the body varies enormously between individuals. However, it is interesting to note that the body usually contains a far greater amount of energy in the form of **fat** than in the other forms of storage. Although the store of energy as **protein** is large, protein is not normally used as an energy source. **Carbohydrate** energy is very important to the body, especially during exercise, but the store of energy in this form is

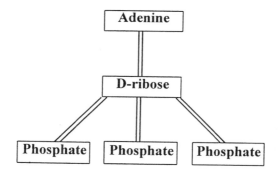

Figure 6.1 Adenosine triphosphate (ATP)

Table 6.1 Typical amount of energy stored within the different fuel stores in a 70 kg human body (~15% body fat)

Fuel store	Energy (kJ)	Percentage of total
Carbohydrate (muscle, liver, blood)	5000	0.8
Fat (adipose, muscle, blood)	500 000	83
Protein	100 000	16
Adenosine triphosphate, ATP	10	0.002

far smaller than the store of energy as fat. And even compared with the store of energy as carbohydrate, the ATP energy store is minute.

Mechanical energy

Only a relatively small amount of chemical energy is converted to useful energy in the human body at rest, with the remainder being converted to heat. This is why the human body is referred to as being 'inefficient' (discussed further in the 'Efficiency' section later in the chapter). Depending upon the type of activity, about 70% of the chemical energy may be converted to heat during physical activity.

However, once the chemical energy is converted to mechanical energy (through the conversion of chemical energy in the muscles), the body is able to perform various physical tasks. Such tasks might range from **involuntary** ones such as contraction of the heart muscle or the respiratory muscles, to **voluntary** ones such as writing or sprinting for the bus. The ability of the body to convert chemical energy into mechanical energy is why it is sometimes referred to as a machine, although not a particularly efficient one! Most of the mechanical energy generated by the body can be quantified (i.e. put in a numerical form) by examining the amount of useful work done. However, because the body is not always designed optimally to perform mechanical tasks, some chemical energy is wasted.

In the study of sport and exercise, it is possible to determine the amount of energy expended by an individual to perform a given task. The energy expended will be the sum of the useful work done plus energy lost as heat and energy lost in other forms (e.g. sound). The total **energy expenditure** is usually measured in an indirect way through the examination of gas exchange at the mouth (see Chapter 2 for further details). Suppose, for example, an individual

Table 6.2 Amounts of various foods that provide 1000 kJ

59 g of cheddar cheese
61 g of corn flakes
62 g of rice crispies
100 g of white bread
150 g of chicken meat
225 g of cod
260 g of banana
270 g of potato
400 g of whole milk

expends 1000 kJ in order to run 5 km. It is interesting to examine what this amount of energy relates to in terms of an amount of chemical energy in the form of food (see Table 6.2). The ability to quantify energy is very useful in sport and exercise, especially when relating physical activity to nutrition.

Kinetic and potential energy

It is important to be able to distinguish between different forms of mechanical energy. An example that is often used to explain two energy types (kinetic and potential) is a water-driven electricity generator. The water is stored behind a dam until it is required, and in this position it has all its energy stored as **potential energy**. When the water is required to turn the water wheel, the dam is opened, and the potential energy due to the water's position under the influence of gravity is converted to **kinetic energy**.

Potential energy is calculated by multiplying the mass (in kg) of an object by the acceleration due to gravity (in m·s^{-2}) acting on the object and the vertical displacement (i.e. height) of the object (in m) (see Equation 6.4). Kinetic energy is calculated by dividing the mass (in kg) of an object by 2, and then multiplying by the square of the velocity (in m·s^{-1}) (see Equation 6.5).

$$\text{Potential energy} = \text{Mass (kg)} \times \text{Gravity (m·s}^{-2}) \times \text{Height (m)}. \tag{6.4}$$

$$\text{Kinetic energy} = \text{Mass/2 (kg)} \times \text{Velocity}^2 \text{ (m·s}^{-1}). \tag{6.5}$$

The transformation of energy between potential and kinetic forms can also be observed when elastic is stretched and then released. As the elastic is stretched it gains potential energy, which is known as strain energy. The maximal gain in strain energy is reached just before the elastic reaches its elastic limit, otherwise known as the **yield point** (beyond this point energy is lost and the elastic takes on plastic properties). If the elastic is then released, potential energy is converted to kinetic energy. The change in length of elastic (m) is proportional to the force applied (N), a relationship known as Hooke's law (see Equation 6.6).

$$\text{Change in length (m)} \propto \text{Force (N)}. \tag{6.6}$$

The gain in strain energy is related to the ability of the elastic to store energy on lengthening (k) and the change in length of the elastic (x) (see Equation 6.7).

$$\text{Strain energy} = \tfrac{1}{2}\,kx. \tag{6.7}$$

The mechanical properties of muscle and connective tissue are very much like those of elastic, except for a damping effect (i.e. a shock absorber effect). In Application 2 at the end of this chapter, the effects of these properties of muscle and connective tissue are examined. Try to investigate the contribution of the elastic properties of the muscle and connective tissue in your own legs by comparing the height jumped with two jumping techniques. First try a vertical jump starting from a crouched position (i.e. with your legs bent). Secondly try

Kinetic energy   Potential energy   Kinetic energy

Figure 6.2 Transformation of energy during a pole-vault take-off

a vertical jump starting from a standing position. You will observe that the height jumped using the second technique is greater due to the elastic contribution of the muscle and connective tissue in the legs. Kangaroos make extremely good use of the elastic properties of the tendons in their legs, to the extent that they become slightly more efficient as their speed increases.

We can find examples of the transformation of energy during sport and exercise that include both the elastic principle and the change in height principle. A simple example is a pole-vaulter. As the pole-vaulter approaches the take-off, he/she has generated much kinetic energy. All of a sudden, most of the kinetic energy is converted to potential energy as the pole is deformed. As the pole-vaulter is propelled upwards, the potential energy stored within the pole is converted to kinetic energy once again. Once the pole-vaulter is passing over the bar, all the kinetic energy is converted to potential energy again, and the maximum potential energy is gained (see Figure 6.2). See Applications 1 and 2 below for further sport and exercise examples.

Work

Essentially, in terms of physical activity, work is the useful end product of the chemical energy used. The useful work done by the body can be accurately measured in some types of activity. It is when the useful work done by the body is compared with the chemical energy used that a measure of the body's work efficiency is obtained (see the 'Efficiency' section on p. 74).

Work is measured in joules (J) (the same unit as used for energy), and is calculated by

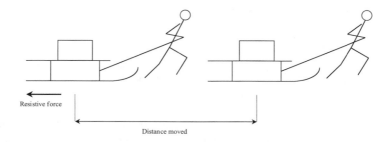

Resistive force

Distance moved

Figure 6.3 Work done in dragging a sledge between two points

multiplying the distance an object is displaced (in metres, m) by the force opposing the displacement (in newtons, N) (see Equation 6.8 and Figure 6.3). Notice the similarity with Equation 6.1.

$$\text{Work done (J)} = \text{Force (N)} \times \text{Displacement (m)}. \tag{6.8}$$

At this point it may be seen that 1 newton metre is the same as 1 joule (i.e. 1 Nm = 1 J).

In many sport and exercise situations it is difficult to assess the amount of useful work done. It was for this reason that a standardised situation was set up in sport and exercise laboratories where the useful work done by an individual could be determined. This is described in detail in Application 4 later in the chapter.

It is very difficult to determine directly the useful work done by a runner, but since running is a very common mode of exercise such a measure would be useful. Various attempts have been made to crudely estimate the work done by a runner. One of these is described below, and another is given in Application 3 later in the chapter.

The potential energy equation (Equation 6.4) may be used for determining work done when running up a gradient. The work done (or useful energy output) is equal to the mass of the runner multiplied by the acceleration due to gravity acting on the runner and the vertical displacement (or gain in height). If the runner has a body mass of 70 kg and the gain in height was 100 m, the work done would be:

$$\begin{aligned}
\text{Useful work done or gain in potential energy} &= \text{Mass} \times \text{Gravity} \times \text{Height} \\
&= 70 \text{ kg} \times 9.81 \text{ m·s}^{-2} \times 100 \text{ m} \\
&= 68\,670 \text{ J or } 68.7 \text{ kJ}.
\end{aligned}$$

The above estimate is very crude, since in addition to the overall gain in height the runner gains and loses height with each stride, and this is not accounted for. It is for this reason that running along a perfectly horizontal road or track at a constant velocity still requires significant work to be done in addition to that required to overcome air resistance.

Power

Often in the study of exercise and sport it is useful to know the amount of work being done per unit time, or, put simply, how quickly work is being done. The rate at which work is being done is usually the factor that determines success in exercise and sport activities, and is termed **power**.

The definition of power is work (in joules, J) per unit time (in seconds, s). The internationally accepted unit of power is the **watt** (W) (see Equation 6.9).

$$\text{Power (W)} = \text{Work (J)/Time (s)}. \tag{6.9}$$

Since work has the same unit as energy, the joule, power is also energy per unit time. This is why the light energy a light bulb emits is also measured in watts.

Since the power generated during exercise is often the main determinant of **performance**, the power that an athlete can achieve is often used as a measure of performance. In studies

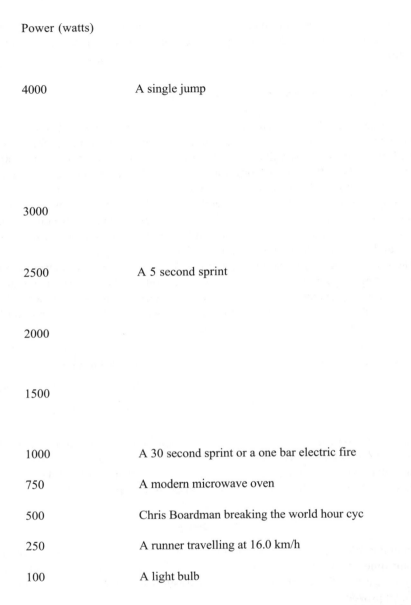

Power (watts)

4000	A single jump
3000	
2500	A 5 second sprint
2000	
1500	
1000	A 30 second sprint or a one bar electric fire
750	A modern microwave oven
500	Chris Boardman breaking the world hour cyc
250	A runner travelling at 16.0 km/h
100	A light bulb

Figure 6.4 Power maintained during a range of physical activities in comparison with other familiar power producers

of sport and exercise, reference is often made to the maximum power generated during an exercise test, or the power sustained over a set period of time. Sometimes individuals are required to exercise at a set power for as long as possible to examine the **capacity** for exercise. Again, this measure is often used as an indicator of performance. Figure 6.4 shows some examples of power maintained during a range of physical activities in comparison with other familiar power producers.

Figure 6.5 Chris Boardman training in the laboratory (Reproduced by kind permission of Mr Bob Seago and the University of Brighton)

It is often useful to compare power produced by individuals during physical activities to get a measure of the differences between individuals or the demands of their specialist athletic events. For example, it is interesting to consider that most individuals could maintain a power output on a bicycle of only about 100 watts for an hour, whereas the cyclist Chris Boardman maintained a power of 450 watts for an hour when he broke the hour record in 1997 (see Figure 6.5).

The ability of an athlete to produce instantaneous power, or power over a very short period of time, is important for certain athletic events. This ability would be useful for events involving jumping, throwing, or accelerating very quickly (e.g. a sprint start). It is possible to estimate power production by asking an athlete to perform a vertical jump. The vertical jump test is often referred to as a Sargeant jump, after a medic, Dr Dudley Sargeant, who practised at Harvard University in the 1870s (Sargeant 1906). Providing the body mass of the athlete is known, the height achieved by the athlete allows determination of the gain in potential energy, and therefore the energy produced. So, if the athlete's body mass is 70 kg and the height jumped is 0.80 m, using Equation 6.4 the gain in potential energy may be calculated as:

$$\text{Potential energy} = \text{Mass} \times \text{Gravity} \times \text{Height}$$
$$= 70 \text{ kg} \times 9.81 \text{ m·s}^{-2} \times 0.80 \text{ m}$$
$$= 549.36 \text{ J.}$$

As long as the time over which the energy is produced is known, the power can be calculated (using Equation 6.9). If the time from the start of muscular force production to take-off is 0.4 s, the power produced is:

$$\text{Power} = \text{Energy or work (J)/Time (s)} = 549.36 \text{ J}/0.4 \text{ s} = 1373.4 \text{ W.}$$

Power produced over a short period of time may be determined by dividing the gain in potential energy when sprinting up steps by the time taken to climb the steps. This step test is referred to as the Margaria step test, after Rudolf Margaria (Margaria 1966). Such short duration power production is required for a short sprint in a team game for example. If an athlete has a body mass of 80 kg, the height of the steps is 2.0 m, and the athlete took 1.1 s to climb the steps, power would be calculated as follows:

$$\text{Gain in potential energy (J)} = \text{Mass} \times \text{Gravity} \times \text{Height}$$
$$= 80 \text{ kg} \times 9.81 \text{ m·s}^{-2} \times 2.0 \text{ m}$$
$$= 1569.6 \text{ J.}$$

The power is therefore:

$$\text{Power (W)} = \text{Gain in potential energy (J)/Time (s)}$$
$$= 1569.6/1.1$$
$$= 1426.9 \text{ W.}$$

A test has also been developed in which it is possible to measure the power produced over longer periods of time, using a **cycle ergometer**. This test is known as the Wingate test, after the Wingate Institute in Israel, where it was developed in the 1970s. The Wingate test is described in Application 5 below, and a typical power curve from a Wingate test is given in Figure 6.8.

Efficiency

It is very useful to compare the rate at which the body is using energy (energy input, usually referred to as energy expenditure) with the rate at which useful work is being done (energy output). Since energy input and useful energy output are being considered, this comparison is known as **efficiency**. Efficiency (%) is energy output (J) divided by energy input (J) multiplied by 100 (see Equation 6.10).

Efficiency (%) = Energy output (J)/Energy input (J) × 100.   (6.10)

An example of the determination of efficiency during physical activity will now be provided. On a cycle ergometer in the laboratory, it is possible to measure the amount of energy being utilised, or the energy used for the exercise, and the amount of useful energy produced, or work done during the exercise. Over 1 minute, for example, it may take 20 000 J of energy to exercise steadily on a cycle ergometer whilst producing 3600 J of energy as useful work. Using Equation 6.10, the efficiency of the human body in this situation is therefore:

Efficiency (%)= Energy output (J)/Energy input (J) × 100

= 3600/20 000 × 100

= 18%.

At this point let us look more closely at how efficiency is defined. Notice until now efficiency has been referred to simply as the relationship between the energy expenditure of the body (energy input) and the useful work done (energy output) during the exercise. However, it is important to be more specific about the energy expenditure by the body during exercise.

Gross efficiency refers to the energy expenditure during exercise as being the energy for the activity plus the energy for other needs of the body at that time, whereas net efficiency refers to the energy expenditure for the exercise alone, without the energy for other needs of the body. The question then arises, how is it possible to separate what energy is used for the exercise and what is used for other requirements of the body? The answer is that the energy utilisation at rest must be measured to establish the energy used for requirements other than the exercise. If in the above example it is assumed that of the 20 000 J used during exercise, 5000 J are used to support resting energy expenditure, the net efficiency is:

Efficiency (%)   = 3600/15 000 × 100

= 24%.

Some researchers even suggest that the energy cost of cycling without a load should be subtracted from the overall energy expenditure during exercise, which provides a measure of work efficiency (Whipp and Wasserman 1969). During cycling work efficiency is reported to be about 30%.

Interestingly, during running exercise, humans are more efficient than when cycling due to the energy stored with each stride within the muscles and connective tissues in the form of potential energy (see Application 2). Although very different values for running efficiency are given in the literature, it is likely that a value in the region of 50% is realistic.

APPLICATION OF SCIENCE TO EXERCISE AND SPORT

1 Energy transformation during the high jump

During locomotion, the human body is fortunately good at changing kinetic energy into potential energy, and then releasing this energy kinetically when required. In fact, any physical activities that make use of the elastic properties of the muscle and connective tissue are making use of the muscles' ability to convert kinetic energy into potential energy. As an extreme example of this phenomenon, think of the high jumper making an approach to the bar. As the jumper reaches the bar, some of the kinetic energy generated on the run-up is converted to potential energy as the jumper crouches down and the take-off leg bends slightly before take-off. At take-off, all the potential energy stored within the muscles and connective tissue is reconverted to kinetic energy, which propels the jumper vertically over the bar (see Figure 6.6). The vertical acceleration is also aided by a very rapid transformation of chemical energy to mechanical energy at the same time.

2 Energy conversion during running

A further important example of the storage of potential energy occurs during running. As the foot of a runner strikes the ground, large impact forces (sometimes over three times body weight) are absorbed by the muscle and connective tissue in the leg. Fortunately, some of the kinetic energy is converted into potential energy and stored within the muscles and connective tissue. As the hip, knee and ankle joints extend, the potential energy is converted to kinetic energy once again. This ability of the muscles and connective tissue to convert kinetic energy to potential energy for later use results in the use of less chemical energy. Consequently, physical activities which make use of this phenomenon are known to be more efficient (see the 'Efficiency' section above).

3 Estimating work during running in the laboratory

Using a treadmill in a laboratory, it is possible to estimate the work done by a runner when running on a level gradient. A harness is attached round the waist of the runner and then a known mass is added over a pulley at the back of the treadmill. The work of the runner is

Kinetic energy   Potential energy   Kinetic energy

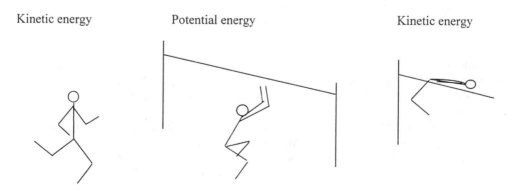

Figure 6.6 Transformation of energy during a high-jump take-off

calculated by determining the product of the force the runner is moving against and the displacement in relation to a fixed point on the treadmill belt (i.e. distance). If we know that the mass added is 5 kg and the distance travelled is 1 km, then, using Equation 6.8:

Work done   = Force × Displacement

and since Force = Mass × Gravity

Work done   = Mass × Gravity × Distance

$$= 5 \text{ kg} \times 9.81 \text{ m·s}^{-2} \times 1000 \text{ m}$$

$$= 49\,050 \text{ J or 49 kJ.}$$

As mentioned earlier, this is a crude method of estimating the work done by the runner. This is because it is assumed in this calculation that the runner does not do work to move in a vertical plane against the acceleration due to gravity. Simply by observing a runner it can be seen that motion does not occur just in a horizontal direction.

4 Measuring work in the laboratory with a cycle ergometer

The cycle ergometer (see Figure 6.7) is a very useful piece of equipment, with which the work done by an individual can be easily quantified.

Just as in the definition of work presented earlier in the chapter (i.e. force × displacement), the assessment of useful work is made by knowing how far the individual is moving the ergometer (i.e. displacement) and multiplying this by the opposing force. The question is, how are the distance and the opposing force calculated on the cycle ergometer? The answer is:

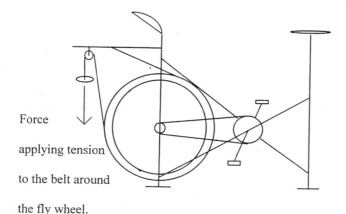

Force

applying tension

to the belt around

the fly wheel.

Work is calculated from the product of the distance the fly wheel moves and the force it

moves against.

Figure 6.7 Work done in cycling on an ergometer

Distance = Number of pedal revolutions × Distance fly wheel moves each revolution (m);

Force = Mass on weight pan (kg) × Acceleration due to gravity (m·s^{-2}).

In the case of cycle ergometers made by a company called Monark, a very common ergometer in physiology laboratories, the distance the fly wheel moves for each revolution of the pedals is set at 6 m. The acceleration due to gravity is constant at 9.81 m·s^{-2}. So the only factors that change are the number of pedal revolutions and the mass on the weight pan. Let us assume for the following example that the pedals are turned at a rate of 60 revolutions per minute (rpm), and that the mass on the weight pan is 1.0 kg.

Work (J) = Force (N) × Distance (m)

$\qquad$ = Mass (kg) × Acceleration (m·s^{-2}) × Revs × Distance per rev (m)

$\qquad$ = $1.0 \times 9.81 \times 60 \times 6$

$\qquad$ = 9.81×360

$\qquad$ = 3531.6 J (in a period of 1 minute, or 3.5 kJ·min^{-1}).

Since in the above example the number of pedal revolutions in a minute was considered, the result was work done in a minute, rather than work done over any other time period. Whilst we usually quantify pedal revolutions over a minute period (i.e. revolutions per minute), the amount of work done is usually calculated over the period that the work takes place. So it would be more useful, in terms of calculating work done, to consider the total number of pedal revolutions during the work period. However, there are some situations when the work performed over a period of time is in fact the desired figure, as you have seen in the section on power above, and can see in the next application.

Since the cycle ergometer represents a real exercise situation very effectively, it will not be surprising to learn that much of the research in the area of exercise physiology has been performed using this type of activity. It is far more difficult to assess the useful work done during other types of exercise activity. For example, although attempts have been made to do so during running, most of these estimate work done in a fairly crude way. This lack of success is due either to incorrect assumptions when assessing work done on motorised treadmills, or to the use of non-motorised treadmills which are unnatural to run on.

5 Determining power output during cycling

Since work done during cycling can be measured, it is relatively simple to then determine power during cycling. Power is the work done per unit time, where time is in seconds. Let us take the value for work done calculated in the previous example (i.e. 3531.6 J). This was performed over a 1 minute period (i.e. 60 s), hence using Equation 6.9:

Power (W) = Work (J)/Time (s)

$\qquad$ = 3531.6/60

$\qquad$ = 58.9 W.

Work done each second (i.e. power) can be determined relatively easily using the Monark cycle ergometer. This concept has been put to good use in the Wingate test mentioned earlier in the chapter. As seen in Application 4, four variables influence the work done, and therefore the power output:

- the mass placed on the weight pan of the ergometer;
- the acceleration due to gravity acting on the mass;
- the distance travelled by the fly wheel for each revolution of the pedals;
- the number of revolutions of the pedals per unit time (minute or second).

The Wingate test is based upon keeping three of the four possible variables mentioned above constant. The variables kept constant are the mass placed on the weight pan of the ergometer, the acceleration due to gravity acting on the mass, and the distance travelled by the fly wheel for each revolution of the pedals. The only variable during the test, and hence the one that will determine the power, is the rate at which the pedals are turned (revolutions per second). Since it is relatively simple to measure the revolutions per second, the power produced at intervals throughout a test can be easily calculated.

Table 6.3 Data collected during a Wingate test

Time (s)	Revolutions per second
1	2.40
2	3.40
3	3.70
4	3.60
5	3.40
6	3.20
7	3.10
8	2.90
9	2.60
10	2.40
11	2.40
12	2.30
13	2.20
14	2.15
15	2.10
16	2.07
17	2.04
18	2.02
19	2.01
20	2.00
21	2.00
22	1.98
23	1.96
24	1.93
25	1.92
26	1.91
27	1.92
28	1.88
29	1.89
30	1.88

For a Wingate test, a participant is asked to pedal as fast as possible for 30 seconds. Prior to the test, the mass placed on the ergometer must be determined. The mass must be such that the participant is not trying to pedal faster than is possible at the start of the test when fresh, but also is not having to stop pedalling at the end of the test when fatigued. Guidelines have been established so that the appropriate mass is used in this regard (see Bird and Davidson 1997: 80).

The data shown in Table 6.3 were collected from a participant during a Wingate test in which the mass placed on the ergometer was 5.0 kg. The acceleration due to gravity was 9.81 m·s^{-2} and the distance travelled by the fly wheel for each revolution of the pedals was 6.0 m.

For the first time point, power is calculated as follows:

$$\text{Power} = \text{Mass} \times \text{Acceleration} \times \text{Distance per rev} \times \text{Revs per second (rps)}$$

$$= 5.0 \text{ kg} \times 9.81 \text{ m·s}^{-2} \times 6.0 \text{ m} \times \text{rps}$$

$$= 294.3 \times 2.40.$$

So for the first second:

Power = 706 W.

Figure 6.8 shows a graph of power against time for the data given in Table 6.3. This is a typical power trace for a Wingate test. It can be seen that power increases quickly in the first 2 seconds, and then reaches a peak at about 6 seconds, after which there is a steady decline. Such a trace provides useful information when studying the energy systems used during short-term exercise.

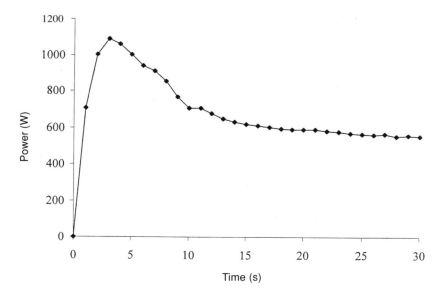

Figure 6.8 A graph of power against time as recorded during a Wingate test

ACTION POINTS

1 Work out the power produced by an athlete exercising on a cycle ergometer against a force of 30 N, and covering 6 m each second.

2 Compare the power that the athlete in the above example is producing with that produced by a 100 W light bulb. Would the athlete be able to generate enough power to light the bulb?

3 Assuming that the athlete is solely dependent upon carbohydrate for energy provision, and the total energy contained within the carbohydrate stores was 360 kJ, how long (in seconds) could the athlete continue to produce the power calculated in point 1 above?

Conclusion

Energy provision is of critical importance for physical activity. It is useful to be able to quantify energy expenditure and the resulting work done. In the study of sport and exercise, energy expenditure is routinely measured, and related to mechanical work output. The relationship between energy expenditure and work output is known as efficiency. It is also useful to quantify the rate at which work is being done. Work rate is referred to as power, and is of great importance in many sporting events. Power output is often closely related to exercise performance. Figure 6.9 illustrates the relationship between energy, work, power and efficiency.

KEY POINTS

- The human body obtains all its energy through food.
- The SI unit for energy is a joule (J), which is the amount of energy required to move the point of application of a force of 1 N through a distance of 1 m.

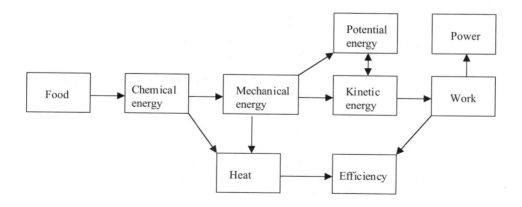

Figure 6.9 Schematic of the relationship between energy, work, power and efficiency

- An additional unit of energy is the kilocalorie (kcal). A kilocalorie is the amount of energy required to heat 1 kg of water by 1°C under standard conditions.
- There are about 4200 J (or 4.2 kJ) to a kcal.
- The body consumes and stores energy in a chemical form.
- Adenosine triphosphate (ATP) is the only form of energy which the body can use directly.
- ATP stores will allow exercise to continue for only about 2 seconds, unless they are replenished.
- About 70% of the chemical energy stored in the body is converted to heat during physical activity, and less than 30% is converted to useful work.
- The elastic properties of the muscles and connective tissue allow the body to convert kinetic energy into potential energy.
- In terms of physical activity, work is the useful end product of the chemical energy used.
- Work (in joules) is calculated by multiplying the distance an object is moved (in metres) by the force opposing the movement (in newtons).
- The rate at which work is being done is important in many sporting events.
- Power (in watts) is calculated by dividing work (in joules) by the time (in seconds).
- Efficiency (in %) is calculated by dividing work done (in joules) by energy expenditure (in joules), and then multiplying by 100.
- Running is a more efficient form of exercise than cycling because of the relatively large storage of potential energy within the muscles and connective tissue during each stride.

Bibliography

Bird, S. and Davidson, R. (1997) *Physiological Testing Guidelines*, 3rd edition. Leeds: British Association of Sport and Exercise Sciences.

Brooks, G.A., Fahey, T.D. and White, T.P. (1995) *Exercise Physiology: Human Bioenergetics and its Applications*. London: Mayfield Press.

Margaria, R. (1966) Measurement of muscular power in man. *Journal of Applied Physiology* 21: 1662.

Maughan, R.J., Gleeson, M. and Greenhaff, P.L. (1997) *Biochemistry of Exercise and Training*. Oxford: Oxford University Press.

McArdle, W.D., Katch, F.I. and Katch, V.L. (1996) *Exercise Physiology: Energy, Nutrition and Human Performance*. London: Lea and Febiger.

Newsholme, E.A., Leech, A. and Duester, G. (1994) *Keep on Running: The Science of Training and Performance*. Chichester: John Wiley & Sons.

Sargeant, D.A. (1906) *Physical Education*. Boston, USA: Ginn and Co.

Whipp, B.J. and Wasserman, K. (1969) Efficiency of muscular work. *Journal of Applied Physiology* 26(5): 644–648.

Winter, E.M. (1990) Assessing exercise performance – the development of terms. *British Journal of Physical Education Research Supplement* 6: 3–5.

Further reading

Bar-Or, O., Dotan, R. and Inbar, O. (1977) A 30-second all-out ergometric test – its reliability and validity for anaerobic capacity. *Israel Journal of Medical Sciences* 13: 329.

Winter, E.M. (1991a) Assessing exercise performance – maximal exercise I. *British Journal of Physical Education Research Supplement* 9: 12–14.

Winter, E.M. (1991b) Assessing exercise performance – maximal exercise II. *British Journal of Physical Education Research Supplement* 10: 13–18.

7

ELECTRICITY

AIMS OF THE CHAPTER

This chapter aims to provide an understanding of the scientific principles of electricity which are relevant to the theory which underpins sport and exercise. After reading this chapter you should be able to:

- define electricity;
- understand electrical charge;
- understand the movement of electrons;
- state the principle of electrical energy;
- state Coulomb's and Ohm's law;
- apply the scientific principles of electricity to sport and exercise.

Introduction

The topic of electricity is very important not only in sport and exercise but also in our daily lives. The majority of the population are dependent on electricity for their work, social and leisure pursuits. The study and investigation of electricity has a long history which amounts to a classic *Who's Who* of scientists. Included in this list are many scientists who have given their names to terms that we use today, such as Count Alessandro Volta who gave us the term volt, Georg Simon Ohm who gave us the term ohm, and Nikola Tesla who invented the Tesla coil. In sport and exercise many measurements depend upon electricity; for example the electrical activity of the heart is measured in the form of an electrocardiogram (ECG).

Scientific principles of electricity

The atomic model of electrons, protons and neutrons states that in an atom there are electrons (which possess a negative charge), protons (which possess a positive charge) and neutrons (which possess no charge and are therefore termed neutral). As the electrical charges of one proton and one electron are equal in strength, when the number of electrons and protons is equal there is no overall charge, and the molecule is said to be neutral. Protons and neutrons in the nucleus are held together tightly, but some of the electrons orbiting around the nucleus are held very loosely. Consequently, the electrons can move from one atom to another. It was George Johnstone Stoney in 1874, who proposed the model of the charge carrier as a particle.

Static electricity

As discussed in Chapter 4, solid materials classified as conductors (e.g. most metals) have loosely held electrons, which can move easily. Conversely, materials classified as insulators (e.g. plastic and glass) have tightly arranged electrons so the electrons cannot move as freely. However, if two materials are rubbed together there can be movement of electrons from one to the other, even if both are insulators. In fact, the Greeks observed this when amber was rubbed against cloth, wool or fur, some 2000 years ago; it was because of this that George Johnstone Stoney used the word 'electron', from the Greek word *elektron*, meaning amber.

Even now, it is not entirely clear how the movement of electrons occurs. Movement is not thought to be due to the friction between the two materials *per se*. At present it is thought that the mere contact between the materials may cause the electrons to move. The result of two different materials coming into contact with one another (e.g. rubbing a balloon on a woollen jumper) creates **static electricity** due to an imbalance in the positive and negative charges. The other name often used for static electricity is electrostatics, which is the study of electric charges at rest.

The gold-leaf electroscope (see Figure 7.1) is an instrument used in some GCSE science work to demonstrate electrostatics and visually show the effects of a small charge flowing through a thin piece of gold leaf. The electroscope is a square container that contains a metal cap, a Perspex plug, and a metal rod to which a gold leaf, or pointer, is attached. When a positively charged rod is brought towards the surface of the negatively charged metal cap,

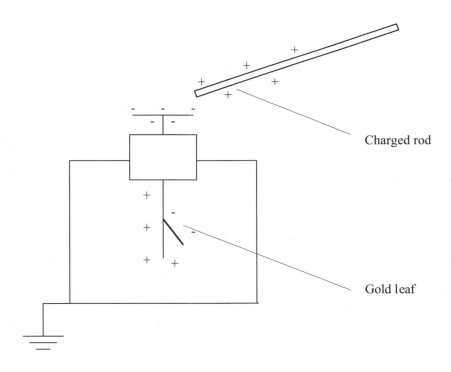

Figure 7.1 A gold-leaf electroscope showing the effects of a positive potential

the potential of the cap is raised. The charge flows on to the leaf and an equal but opposite charge is created inside the case, so that the leaf is deflected. Thus, the deflection of the leaf is approximately equal to the electrical potential.

The adage that opposites attract is related to charges. A proton (positive charge) and an electron (negative charge) will be attracted to one another, but like charges (e.g. two protons) will repel one another. These attraction forces are clearly demonstrated through the example of a balloon and a woollen jumper. When the balloon is rubbed on the jumper it acquires extra electrons and therefore an increasingly negative charge. If the balloon is then placed on a third object, such as a wall, it will stick to the wall due to an attraction between the positive charges in the wall and the negative charges on the balloon. If the object happens to be a good conductor, the negative charges (electrons) in it will move as far away from the balloon as possible. However, the balloon will still stick to the conductor because of the attraction of the positive charges in the latter. Another common example is the spark or small electric shock which is sometimes experienced when you walk across a rug and then brush another person's hand. When you walk across the rug a large number of electrons are transferred from it to you (i.e. you gain increased negative charge). When you touch the other person, they act as a conductor, and the electrons move rapidly from you to them, creating the spark or shock. You will recall from Chapter 4 that the human body is a good conductor because of the large amount of water in its composition. However, it should be stressed that water *per se* is not the prime reason for conduction within the body; rather it is the dissolved ions, such as sodium and chloride, that assist the flow of the charge.

When two objects are rubbed together one becomes negatively charged and one becomes positively charged. To simplify the decision about which objects become negative and which positive, scientists have formulated a rank order list (Figure 7.2). Thus whichever object appears higher along the scale gives up electrons and becomes more positively charged; hence the other object (lower along the scale) will become more negatively charged. This list is known as the triboelectric series.

The imbalance of positive and negative charges, which leads to static electricity, is merely the movement of electrons from one place to another. There is no creation or destruction of electrons, hence the net electric charge stays the same. This concept is called the principle of conservation of charge, and is similar to the conservation of energy principle discussed in Chapter 6.

Coulomb's law

The electric charge of a proton or an electron is measured in coulombs (C). A single positive or negative electrical charge is equal to 1.6×10^{-19} C. Although smaller fractions of charged particles have been found (known as 'quarks'), experimentally it has been difficult to isolate

Human hand Glass Human hair Nylon Wool Fur   Silk   Paper Cotton Hard rubber Polyester   Polyvinylchloride plastic

High   ⟵——————————————————⟶   Low

Figure 7.2 The triboelectric series

them. Therefore, at present all charged particles (negative or positive) are integer multiples of a 'quantised' charge.

The invisible electrical field created around objects is a product of both the positively and negatively charged objects. The force of the field around the objects is dependent on the amount of the charge, the distance between the objects and the shape of the objects. Because of the complexity of the situation, electrical charge is often simplified by treating the charged protons and electrons as a 'point source'. The point source applies to charged objects that are much smaller than the distance between them.

Charles Coulomb, in 1785, was the first to describe the strength of electric fields. Coulomb found that the electrical force varied directly with the product of the charges. This means that the bigger the charge, the stronger the field. He also stated that the field varied inversely with the distance between the charges. In other words, the greater the distance, the weaker the field. These experimental relationships are formulated in what is known as Coulomb's law:

$$F = Kq_1q_2/d^2,$$ \hfill (7.1)

where F is electric force (N), q_1 and q_2 are the two charges (C), d is the distance between the two charges (m), and K is Coulomb's constant, equal to 9×10^9 Nm^2·C^{-2}.

Therefore, from Equation 7.1, if the distance between two charged particles is doubled, the force is reduced by three-quarters. If the charge is doubled, the force is increased by a factor of four.

The discussion so far has dealt only with electric charges at rest (static electricity or electrostatics). Although static electricity and current electricity are the same phenomena, and give the same effects, the former deals with very small charges. In current electricity the charge is much greater, and this is the type of electricity with which we come into contact every day.

Current electricity

When electricity passes through a material such as a wire, electric current can be described as flowing through the material. However, the electric current is more accurately described as an organised 'drift' of electrons. The size of the current is the rate at which the electrons flow. Hence, electrons will be attracted to the positive charge within a wire (or any other material in which the current is travelling) whilst being repelled at the opposite end of the wire because of the negative charge (remember like charges repel). Even though electrons within a wire are travelling at many thousands of metres per second, the potential difference between two points imposes an average overall velocity of a few centimetres per second upon the high speed of the electrons. It is incorrect to think that a particular electron has to travel from one end of a wire to the other. It is perhaps easier to relate to the concept of 'drift' for electrons within materials. The important effect is more of a 'falling domino' scene, whereby one electron will invade the space of the nearest electron and repel it, leading to that electron then moving into the space of the next electron and so on. This domino effect results in the electrical energy travelling at the speed of light (3×10^9 m·s^{-1}). Therefore, electricity should be considered as a form of kinetic energy when it is flowing in a wire.

Electric current is measured in amperes. One ampere is equal to 1 coulomb per second:

$1 \text{ A} = 1 \text{ C·s}^{-1}$.

A more detailed and technical definition of the ampere, often found in physics textbooks, is:

> The electric current which, when flowing through two straight parallel conductors of infinite length and negligible circular section and one metre apart, causes a force between the wires of 2×10^{-7} newtons per metre.
>
> (Duncan 1994: 272)

An understanding of electricity in the context of atoms and matter should aid an understanding of energy and forces. Electricity, by virtue of the position of the protons and electrons, can be described as possessing potential energy in much the same way that a battery can store energy. This potential is associated with another known unit of electricity, the volt. The volt is a measure of the difference in potential energy between two points, known as the potential difference. Potential difference is very similar in concept to the potential gradient of liquids discussed in Chapter 3. A potential gradient allows a liquid to flow from one point to another providing there is a difference in the pressure at the entrance and exit points. Electrons tend to flow to the point of least potential energy. Electricity can be thought of as a method of transferring energy from one location to another. Another way to look upon the voltage, or potential difference, is in terms of the driving force. The volt is the equivalent of 1 joule (J) per coulomb (C). The volt can be defined as:

> ... the potential difference such that one ampere of current will be driven through a resistance of one ohm.
>
> (Duncan 1994)

In the above definition, the term resistance was used. Resistance is measured in ohms, after Georg Ohm. Ohms are signified by the Greek capital letter omega, Ω. The resistance of matter is a function of the type of the material, and of the length and thickness of the material. The resistance to electricity will therefore impede the flow of electrons. Ohm's law (Equation 7.2) describes how current, voltage and resistance relate to one another:

$$I = V/R, \tag{7.2}$$

where I is electric current, V is potential difference, and R is the resistance of the conductor. The formula can also be rearranged to give the resistance of a conductor:

$$R = V/I. \tag{7.3}$$

The three main factors in electricity mentioned so far – potential difference (voltage), current and resistance – are illustrated in Figure 7.3. This diagram shows the form of an electrical circuit. Notice the direction of the arrows in Figure 7.3: the arrows are deliberately pointing away from the positive terminal of the battery towards the negative terminal. However, from the information given so far we have understood that direction of the electric

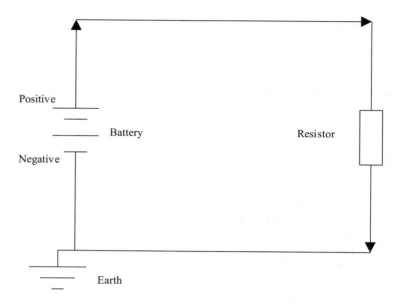

Figure 7.3 An electric current circuit

current is from negative to positive. This apparent contradiction can be explained as follows. Historically, the direction of electric current was thought to be from positive to negative, as if the positive charge were responsible for the flow of charge. Therefore, in all schematics of electric current the direction of the arrow for flow was from positive to negative. However, it is now known that the flow is from negative to positive and that electrons are responsible for this flow of charge. Since the effect of the negative charge moving to the positive is exactly the same as the positive charge moving to the negative, this apparent contradiction should not cause a problem.

Since electricity is a force that transfers potential energy to kinetic energy, it is possible to express electricity in units of energy, such as joules, or units of power, such as watts. As stated in Chapter 6, the joule is the SI unit of energy or work, and 1 joule of work is done when a force of 1 newton acts through a distance of 1 metre. The watt is the SI unit of power, and 1 watt is 1 joule of work done per second. In electrical terms it is common to come across the term kilowatt-hour (kWh). This is a non-SI unit of energy, and is defined as the energy consumed in 1 hour by an electrical appliance working at a power of 1 kilowatt (1000 watts), such that 1 kWh equals 3.6×10^6 joules. If the energy change, or work done, is represented by W (take care not to confuse this symbol with the abbreviation W for watts) and measured in joules (recall that $1\ V = 1\ J{\cdot}C^{-1}$) then:

$$W = Q \times V, \tag{7.4}$$

where Q is coulombs and V is potential difference in volts. Since $Q = I \times t$:

$$W = V \times I \times t, \tag{7.5}$$

where I is current (in amperes) and t is time (in seconds).

To be translated into power (*P*), the formula for work (*W)* must now incorporate the factor of time because by definition power is the rate of work per unit time. Therefore, the formula now becomes:

$$P = (V \times I \times t)/t, \tag{7.6}$$

or

$$P = V \times I.$$

Electric current can be either what is known as direct current (d.c.) or alternating current (a.c.). A torch battery or a car battery are examples of a direct current, whilst mains electricity is an example of an alternating current. The mains electricity which is delivered to our homes has a potential of approximately 240 V. If the equation $P = V \times I$ is rearranged to $I = P/V$, then we can calculate that a typical 120 W lamp running off 200 V would draw a current of 0.66 A.

APPLICATION OF SCIENCE TO EXERCISE AND SPORT

1 A defibrillator

A defibrillator is basically a large capacitor, which delivers an electric shock to the heart. A capacitor is designed for storing electric charge so a defibrillator therefore has an electric potential or potential voltage. A conducting pathway must be made available for a capacitor to discharge, thereby releasing the stored electrical energy. By placing what are known as the paddles of the defibrillator on the chest of a patient, the pathway is completed. A switch in the circuit is then closed, allowing the capacitor to discharge the current in a few milliseconds. The switch is then opened, breaking the pathway for the electricity, and the defibrillator recharges in preparation for the next occasion. A defibrillator has a potential of about 7500 volts and the charged capacitor has a stored energy of about 400 J.

A defibrillator is used to deliver an electric shock by discharging a high voltage through the heart. The electric shock is used to overcome the heart's irregular rhythm, and if successful allows the heart to beat in a normal rhythm again. A discharge of 200–360 J will usually result in depolarisation of the heart muscle (the myocardium). After an electric shock the heart should be capable of providing its own electrical stimulus, thereby returning to a normal ECG wave formation (see below). When a patient is having a heart attack (i.e. in ventricular fibrillation), the normal electrical impulse of the heart, from the sinoatrial node to the atrioventricular node, through the Purkinje fibres and through the ventricles, does not occur. Hence, the irregular electrical pattern must be disrupted by the action of the defibrillator, and the normal electrical activity restored.

2 Electrocardiogram (ECG)

The electrocardiogram (ECG) is one of the most widely used techniques in both hospitals and exercise physiology laboratories. The ECG is a graphic representation of the electrical forces produced by the heart. An ECG relies on the body's water and salt content to allow the heart's voltage of approximately 1 mV to be detected at the surface of the skin. The heart has a natural rhythm originating from the sinoatrial node. With a three-lead ECG the electrode leads, often labelled as I, II and III, are used to detect both the magnitude and direction of the heart's electrical impulse. Prior to an ECG wave form being determined, the skin has to be prepared with a swab to allow for optimal conductance of the electrical signal. This procedure often includes shaving off any excess body hair and the removal of dead skin cells by vigorous rubbing with an alcohol wipe. Electrodes are then placed on to the skin. A standard placement for lead I is just below the left collar bone (clavicle); lead II is placed on the sixth rib on the left hand side in a vertical line with lead I, and lead III is put on the eighth rib on the right hand side. Leads I, II and III are positive, negative and earth leads respectively. The ECG device is able to convert the electrical signal into what we see as the heart's PQRST complex (see Figure 7.4). The original ECG lead system was devised by a scientist called Einthoven. The system uses what is known as Einthoven's triangle to determine the sum of all the electrical activity within the heart. By assessing the magnitude and direction of the electrical signal to lead I, to lead II and from leads I to II, and by using Pythagoras's theorem and simple geometry, the signal is converted into a wave form of electrical activity of the heart, which can then be interpreted. Heart rate monitors use a similar principle to determine heart rate with the use of two electrodes that form part of the chest strap.

3 Bioelectrical impedance

Bioelectrical impedance analysis (BIA) is a common technique used in body composition assessment. It is based on the fact that electrical flow is passed more easily through fat-free tissues and extracellular water than through fat tissue. The reason for this is the greater content of electrolyte in fat-free tissue. Thus BIA will record a lower resistance for lean tissue than for fat tissue, and hence the impedance of electricity is directly associated with fat tissue.

The protocol used in measuring body fat by the BIA method typically involves requesting a participant not to eat for 4 hours nor to exercise for 12 hours prior to the measurement being taken. Similarly, abstinence from alcohol for 24 hours prior to the test is also requested. The instrument requires the participant to be in the supine position and as relaxed and still as possible. One electrode is placed on the phalangeal-metacarpal joint on the dorsal surface of the right wrist and the second electrode is placed on the medial side of the right wrist. A third

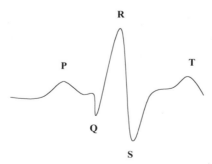

Figure 7.4 An electrocardiogram (ECG) trace showing the PQRST complex of electrical activity

electrode is placed just below the metatarsal arch on the superior side of the right ankle and another on the medial bone of the right ankle. A small, undetectable electrical signal (50 kHz, 800 μA) is passed through the body and a resistance to the current is recorded. The impedance figure plus details of gender, stature, body mass and age are then used to predict density. This figure is then converted to percentage body fat using the Siri equation (Siri 1961).

Although this technique is quick, easy to use and non-invasive, there are a number of confounding factors which can easily distort the true value. The first factor is the participant's hydration level. Numerous studies have shown that the hydration level of the participant alters the recorded value. For example, a deficit of body water will decrease the impedance value and the participant will record a lower body fat percentage. Conversely, a hyperhydrated state will result in an artificially high body fat percentage. The temperature of the skin also affects the predicted body fat score: measurements taken in warmer climates typically result in lower impedance values than measurements taken in colder climates.

The BIA technique is thought to be not as accurate as other body composition techniques, such as skinfolds and hydrodensitometry. In fact, because the BIA method has been validated against skinfold testing it can never be more accurate than skinfold testing itself. However, though it should be used with caution, it does provide another non-invasive technique for predicting percentage body fat (Williams and Bale 1997).

ACTION POINTS

1 How does an insulator differ from a conductor?
2 Why would dissolved ions such as sodium and chloride assist the flow of charge within the human body?
3 Investigate how strong other forces, such as magnetism and gravity, are in comparison with electrical force.
4 Construct a table to show how coulombs, amperes, volts and ohms all relate to one another.
5 Would a thicker and shorter wire have a lower or higher resistance than a wire of the same material but which was thinner and longer?
6 Describe how electrical behaviour is similar to that of liquids.
7 Find some examples of direct and alternating current systems.
8 Review Figure 6.4 in Chapter 6, noting the power outputs of human performance and electrical appliances.
9 Find out the electrical output of the heart and suggest why an ECG signal can be distorted during an exercise test.
10 Electromyograms (EMG) measure skeletal muscle electrical activity. Explain why the concept is similar to that of an ECG instrument.

Conclusion

Electricity is a force, which relates to the atomic model. The study of electricity involves looking at the interactions of electrons. In electrostatics (or static electricity), the attraction of positively and negatively charged protons and electrons is central to the effects shown by the electrical force. The imbalance of positive and negative charges responsible for static electricity is due to the movement of the protons and electrons on an object, which results in a net electrical force. Therefore, energy is neither created nor destroyed, conforming to the principle of conservation of charge.

As an object carries both negative and positive charges it creates an electrical field. The force of this field is dependent upon the amount of the charge, the distance between the objects and the shape of the objects. As a result of research by Charles Coulomb, the strength of electrical fields was found to vary directly with the product of the charges. This finding led to the law known as Coulomb's law.

Electricity can also flow through conducting materials such as metal, and this current is measured in amperes. One ampere is equal to 1 coulomb per second. The driving force or potential difference contributing to this electric current is known as the voltage. To establish the relationship between the electric current and the voltage, the resistance of the material must be known. Resistance is a function of the type of material and its thickness. The work of Georg Ohm led to a law known as Ohm's law which describes the relationship between current, voltage and resistance of a conductor. As electricity is a force, it is possible to express it using units of work or power. The properties of electricity are used in several instruments in the study of exercise and sport.

KEY POINTS

- The effect of the behaviour of electrons has been observed since the time of the Greeks, and since then many great scientists have studied these phenomena. These include Count Alessandro Volta, Benjamin Franklin, Georg Ohm and James Clerk Maxwell.
- Electricity is the flow of electrons through matter.
- Metals are particularly good conductors of electricity due to the fact that some of their outer electrons are not attached to a single atomic nucleus but instead are free to move within the lattice structure of the metal.
- Static electricity is due to an imbalance of positive and negative electric charges.
- An atom that loses electrons acquires a positive charge and an atom that gains electrons acquires a negative charge.
- An ion is a charged particle; a cation is a positively charged ion and an anion is a negatively charged ion.
- Like charges repel whilst unlike charges attract.
- Coulomb's law states that the force between two charges is directly proportional to the product of the charges divided by the square of the distance between them.
- An ampere is equivalent to 1 coulomb per second. This is a measure of the quantity of electricity flowing, i.e. the electric current.
- The volt is the driving force of electricity and provides the potential to drive a current of 1 ampere through a resistance of 1 ohm.
- The resistance is the impedance to flow of electrons and depends on the material, its length and its thickness. The unit of measurement is the ohm.
- Ohm's law states the relationship between voltage, resistance and current. The voltage divided by the resistance equals the current.
- Electricity should be considered as a method of transferring energy.
- Electricity can be expressed in units of power (W).
- In electricity there are four main areas to relate to physiology: the law of conservation

of electrical charge; opposite charges attract; energy is required to separate protons and electrons; and conductors and insulators.

Bibliography

Duncan, T. (1994) *Advanced Physics*, 4th edition. London: John Murray Publishers.

Siri, W.E. (1961) Body composition from fluid spaces and density: analyses of methods. In *Techniques in Measuring Body Composition*, J. Brozek and A. Henschell (eds), pp. 223–244. Washington, DC: National Academy of Sciences/National Research Council.

Williams, C.A. and Bale, P. (1998) Bias and limits of agreement between hydrodensitometry, bioelectrical impedance and skinfold calipers measures of percentage body fat. *European Journal of Applied Physiology* 77: 271–277.

Further reading

Isaacs, A. (1963) *Introducing Science*. Middlesex: Penguin Books (pp. 139–142).

Part 3

SCIENTIFIC TRANSFERABLE SKILLS

8

DATA ANALYSIS

AIMS OF THE CHAPTER

This chapter aims to provide an understanding of the principles of data analysis which may be applied to the study of exercise and sport. After reading this chapter you should be able to:

- illustrate the importance of data analysis in sport and exercise;
- demonstrate the link between data analysis and hypothesis testing;
- demonstrate the link between data analysis and experimental design;
- outline the fundamental principles of scientific data analysis;
- provide examples of data analysis in exercise and sport.

Introduction

Considerable thought and planning usually precedes any scientific investigation, from the stage of isolating the problem, through to designing the study. In this chapter some of the important principles upon which scientific investigations are based will be considered. Without an understanding of the principles of the scientific method, hypothesis testing, and experimental design, it is difficult to consider data analysis. Consideration of the construction of an investigation, from the recognition of the problem to the analysis of the data, will aid the understanding of why certain types of data analysis are utilised. The remainder of the chapter will consider the different types of data that may be collected when studying sport and exercise, and the possibilities for appropriate analysis.

In the study of exercise and sport, it is often necessary to collect information either through measurement or through questionnaires and surveys. Sometimes information is also gathered through observation of individuals and their behaviour. When a scientific investigation takes place, information is usually collected through **direct measurement** (normally producing quantitative data), and this is the type of information that will be considered in this chapter. For further details of the analysis of qualitative data, which may be collected through observation, interviews or questionnaires, see Patton (1980), Silverman (1993), and Derzin and Lincoln (1994, 1998).

The advantage of making a direct measurement is that there should be no influence of the investigator on the data. Thus the data may be collected by different investigators, who would all obtain similar results. Direct measurement ensures a degree of reliability between investigators (inter-investigator reliability) since the data are collected objectively. However, the findings of an investigation of this type are not completely without some subjective influence. First, the design of the study will be influenced by the investigator, which affects the findings. Additionally, interpretation has to take place by the investigator once the data have been collected in order to place the findings in context and draw conclusions.

This chapter will not cover statistical tests in detail; for information on these Hinton (1995) is a good place to start. Also, it should be noted that because data analysis is nowadays primarily performed using widely available computer software, it is important to read this chapter in conjunction with the one on information and communication technology (Chapter 9).

The scientific approach

Many scientists, particularly in the area of exercise and sport, work within an established scientific **paradigm**. The underlying principle is that scientific understanding is advanced as theories are tested. Scientific theories are proposed, in order to be rigorously tested. If the tests prove that the theories are inadequate, the theories should either be adapted to account for the new findings, or discarded.

Scientific investigations are designed in order to test particular aspects of our understanding. The design of an investigation must be such that our understanding can be challenged with a degree of confidence. It is important to recognise that much data analysis is performed in order that a theory can be challenged with a degree of confidence. In this regard, confidence is usually expressed as a probability or percentage. For example, an investigator might accept a probability of 95 times out of 100, or 95%. However, the area of research often dictates acceptable probability levels. For example, in medical research, where lives are at stake, 95% probability may not be acceptable.

The first step in designing a scientific investigation is the formation of testable hypotheses, based upon the identified problem or **research question**. The design of the investigation will then test the hypotheses, and allow focused conclusions to be drawn (see Figure 8.1).

Hypothesis testing

The Collins English dictionary definition for a **hypothesis** is 'a suggested explanation for a group of facts or phenomena, which is accepted as a basis for further verification, or an assumption used in an argument'. A hypothesis is normally broken down into two components: a null hypothesis and an alternative hypothesis. If the example of an investigation that is attempting to determine whether a particular intervention has an effect is considered, the null hypothesis usually states that there is no effect due to the intervention, while the alternative hypothesis states that there is an effect due to the intervention. If the investigators are confident of the direction of effect, they will normally make that explicit in the alternative hypothesis. For example:

- Null hypothesis (H_0): Running spikes do not decrease race times.
- Alternative hypothesis (H_1): Running spikes do decrease race times.

Whether the investigators accept or reject the hypotheses will depend on what is shown by the data analysis. The particular data analysis techniques used will be selected based upon their ability to determine whether the hypotheses should be accepted or rejected. At this

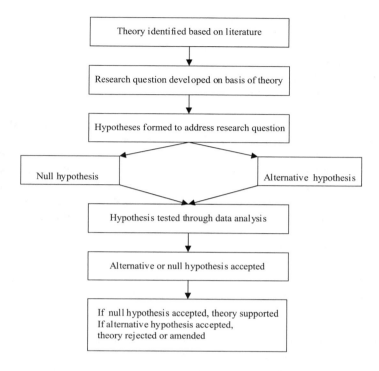

Figure 8.1 Flow diagram of the process of scientific investigation

point a word of caution is required. It must be understood that the hypotheses are only ever rejected or accepted with a degree of confidence. For example, it is never possible to say that on every occasion the null hypothesis is true. Although at first encounter this may not seem adequate for scientific investigations, the concept of a degree of confidence in fact serves a very useful purpose. In every experiment, **random errors** will occur for one reason or another. If a hypothesis were accepted only if every observed result agreed with the hypothesis, the findings of a study would always be influenced by the random errors. Random errors are the deviation in a measurement for no reason other than investigator or participant variation. The degree of confidence is determined by the investigator, and implemented through the chosen statistical test. The degree of confidence (or confidence limit) selected depends on several factors (see the 'Statistical errors and power' section below).

Details on how to include hypotheses in a scientific report can be found in Chapter 10.

Experimental design

When determining the most appropriate experimental design, two key questions should be considered:

1 What is the best design that tests the hypotheses?
2 Does the design lend itself to data analysis?

Both of the above considerations are inextricably linked. The design must allow the hypotheses to be tested, which is only possible if data analysis can be performed. It is the **inferential data analysis** (see 'Inferences from findings' section below) that allows an investigator to claim with confidence that the results within the examined sample may be generalised to the whole population.

Since it is rare in the study of sport and exercise to be able to examine a whole population, it is normal to select a **sample** from the population for examination. However, the way in which the sample is chosen from the population may influence the findings dramatically. It is usually necessary to select a representative sample from the population; failure to do this would result in what is known as a biased sample. A representative sample shares the same characteristics as the population. The key question is then, how is a representative sample selected? Various approaches are implemented to achieve a representative sample, including random sampling and purposeful sampling. Random sampling ensures that each member of the population has an equal chance of being selected for the sample. Purposeful sampling ensures that the characteristics of the sample are representative of the characteristics of the population.

For the remainder of this section a sample will be referred to as a group. It is usual to refer to a group when discussing the participants in relation to the experimental design, but to a sample when discussing the participants in relation to data analysis principles.

The following experimental designs are often employed in the study of exercise and sport:

- a single group which is then compared with a population;
- a control group and an experimental group;
- a single group which is tested prior to and following an intervention;
- a case study performed on a small number of participants who are exposed to an intervention at various time points.

Within each of these broad categories the design of the experiment can be manipulated in a variety of ways, as now considered.

The reason for having a **control group** that is monitored, despite the lack of an intervention, is to determine whether the results in the **experimental group** are due to the intervention alone. If a similar change is observed in the control group, it cannot be claimed with confidence that the change in the experimental group is due to the intervention. In some situations it is possible to have a control and an experimental group that swap over. The control group becomes the experimental group, and the experimental group becomes the control group. This is known as a **cross-over design** (e.g. Kreider *et al.* 1990). Sometimes it is not possible to implement such a design, e.g. if the experimental intervention is not retractable. For example, if it is known that the administration of a drug to the experimental group will have lasting results, it is not possible for that group to become the control group. A cross-over type design is generally considered to be stronger than the simpler control and experimental group design.

A single group is often used to examine the effect of an intervention. This particular method is weaker than a control and experimental group approach. However, such an approach is sometimes unavoidable. For example, if an investigator wishes to test the effect of a drug which is strongly suspected to have only beneficial effects on health, and no negative side-effects, it may be considered unethical to allow one group of patients access to the drug, but not another group. Under this condition, a single group experiment is unavoidable. There are, however, various ways to strengthen the design considerably, even where there is just a single group (see below).

A design that is often considered stronger than a basic control and experimental group is a **repeated measures** design, where the participants all act as their own controls. Usually the participants will be observed prior to an intervention, during the intervention, and following the intervention. If the observed responses return to pre-intervention values when the intervention is removed, it is possible to state with a degree of confidence that the observed change is due to the intervention. If a single group design must be used, e.g. due to ethical considerations as discussed earlier, it is possible to further strengthen the design. Rather than simply having all participants exposed to the intervention at the same time, the application of the intervention could be staggered. With this approach, all participants receive the intervention that is perceived to be beneficial. The benefit of the staggered approach is the opportunity to observe whether any change in the response of the participants who have the intervention administered early is due to the intervention or something else. If a change is also observed in the response of the participants who have not yet been administered the intervention, the observed change is likely to be due to another factor. The other factor may simply be a 'placebo effect' (or Hawthorne effect), which is the term used to describe a change for no apparent reason apart from the fact that the participants think they have been given the intervention.

A similar approach of staggering the application of an intervention is adopted in investigations that use a small group of participants. A **case study** approach often relies on the application of innovative strategies to determine whether any observed change is due to the intervention or not. Inferential analysis of findings is the usual method of determining with a degree of confidence whether any observed effect is due to an intervention. However, inferential data analyses are normally dependent upon relatively large group (sample) sizes. With a case study approach, the number of participants may be anything from one to five or possibly more. To account for the small number of participants, but still allow a statistical approach to the analysis, a multiple baseline is used. The multiple baseline is simply an extension of the repeated measures design referred to above, where a participant acts as their own control. However, with a multiple baseline design the intervention is applied and taken away several times. By adopting this approach, enough information is collected to allow analysis of the findings via inferential statistical tests. Again, such an approach is obviously not possible when an intervention is not retractable. However, if more than one participant is used it is possible to modify the design to stagger the application of the intervention between participants. Once the intervention has been applied, it cannot then be removed without a residual effect. This latter approach is similar to the staggered repeated measures design with larger groups.

Table 8.1 summarises the various experimental designs used in the study of exercise and sport and their relative merits.

From the above discussion it is clear that the design of an experiment cannot be separated from the method of data analysis. Inferential data analysis simply allows the appropriate hypothesis to be accepted with a degree of confidence. The type of analysis will be heavily dependent upon the experimental design. Often, the more complicated the experimental design, the more complicated the analysis – a trap many a student has fallen into! Thus, an understanding of the principles of data analysis will be helpful when choosing an appropriate experimental design.

Table 8.1 Experimental designs used in the study of sport and exercise

Study design	Differences*	Relationships†	Own control‡
Experimental group	✓	✓	
Control and experimental group	✓	✓	
Cross-over	✓	✓	✓
Repeated measures	✓	✓	✓
Multiple baseline	✓	✓	✓
Case study			✓

*The ability to determine an effect or difference between variables.
†The ability to determine relationships between variables.
‡The participants act as their own control.

Analysis principles

Whilst it is not the aim of this chapter to cover particular statistical tests which may be of use to a student of exercise and sport, the intention is to provide enough information to enable the reader to understand the principles underlying all relevant data analysis techniques. In addition, it is intended that the information provided will enable the reader to choose an appropriate statistical test. Several texts provide useful information on specific statistical tests. For a basic level text see Hinton (1995), for an intermediate level see Vincent (1995), and for an advanced level see Howell (1997).

Data analysis is usually divided into two categories: **descriptive data analysis** and **inferential data analysis**. Once data have been collected in their raw form, it is usually necessary to undertake some analysis prior to presentation of them. As the term suggests, descriptive analysis of data enables an observer to obtain important information in order to appreciate the data. Often data presented in their raw form is not the best way to allow an observer to examine the data. Descriptive analysis of the data, however, does not allow an observer to make a judgement about how the findings from the sample under investigation might be generalisable to the population. An inferential analysis of the data is required for this purpose. Prior to either form of analysis, it is necessary to establish what type of data is under investigation.

Type of data

In order to choose the most appropriate analysis technique for either describing data or drawing inferences from them, it is essential to first consider the type of data being dealt with. The type of data is categorised according to the scale of measurement used. The four categories are:

- nominal;
- ordinal;
- interval;
- ratio.

Nominal data are those which are categorised. However, no relationship exists between the categories, and the categories cannot be ordered in any way. For example, human participants may be categorised according to their hair colour: category 1 could be black hair, 2 could be red hair, etc. There is no way of ordering these categories in any meaningful way. We cannot state that category 2 is better or higher than category 1. Perhaps a more meaningful example for sport and exercise would be the sport played by participants. If a study recruited individuals who engaged in running (category 1), football (2), swimming (3), cricket (4), cycling (5), and rowing (6), there would be no way to order such categories. The sport categories could, however, be further categorised into team or individual, or water- and land-based sports.

Ordinal data are those which are categorised, and may also be ordered in some way. However, it is not possible to say that the difference between two categories is the same as that between two other categories. For example, human participants may be categorised according to their perception of how strenuous an exercise task appeared. The degree to which the task was strenuous could be rated between 1 and 5, with 1 being very easy and 5 being very hard. From this information it is possible to claim that the participants in category 2 found the task less strenuous than those in category 4. However, it is not possible to claim that participants in category 4 found the task twice as strenuous compared with those in category 2. So although the categories can be ranked, the numbers ascribed to each category cannot be treated mathematically.

Both nominal and ordinal data are classified as being **category data**. Such data are analysed in a standard way, and statistical tests known as non-parametric tests are normally used to draw inferences from them.

Interval data are measured from a continuous scale of measurement, where the differences between successive increments are uniform. However, such data have no fixed zero reference point; the point that takes the value 0 is arbitrary. Interval data can be treated mathematically, but there are some restrictions, e.g. it is not possible to apply ratio manipulations to such data. An example of interval data is measurements of temperature. Different temperature scales have different fixed zero reference values. For example, the Celsius scale has the freezing point of water as its 0. The boiling point of water (100) is a further fixed reference point. On the Celsius scale each increment between 0 and 100 is known as a degree (°). The kelvin scale has the same magnitude degree increments, but a zero value corresponding to −273° on the Celsius scale. So 0° on the Celsius scale corresponds to 273 on the kelvin scale. Therefore, doubling a figure on the Celsius scale has a different effect from doubling a figure on the kelvin scale.

Ratio data are exactly the same as interval data, except that ratio manipulations of the data are valid. Thus it is possible to state that a score of 40 is twice as high as a score of 20. Most data that are collected on a continuous scale of measurement are ratio data. An example of ratio data would be the height of participants in a study. The zero value is absolute (i.e. no height), and the increments are uniform. For example, the difference between someone who is 165 cm tall and someone who is 185 cm tall is the same as the difference between someone who is 170 cm tall and someone who is 190 cm tall.

Both interval and ratio data are classified as being continuous data. Continuous data are normally analysed in a standard way, and parametric statistical tests are normally used to draw inferences from them (see 'Choosing an appropriate statistical test' below).

For a summary of the four types of data see Table 8.2.

Throughout the remainder of this chapter the emphasis will be on interval and ratio data, since these are the types which normally emanate from the direct measurement techniques employed in laboratory-based investigations of sport and exercise. Fortunately, such data types can also be used within a large range of analysis procedures.

Table 8.2 Characteristics of the different types of data

Type of data	Categorised	Ranked	Addition/ Subtraction	Multiplication/ Division
Nominal	✓			
Ordinal	✓	✓		
Interval	✓	✓	✓	
Ratio	✓	✓	✓	✓

Description of findings

Descriptive analysis is the transformation of data into a 'user friendly' form, allowing the observer to quickly obtain the most important information from them. Normally such information is related to the central point (central tendency) of the data, and the degree of spread (variation) about this centrality. This information can be expressed in a variety of forms, as shown in Table 8.3.

The data in the second column of Table 8.3 are the resting heart rates (HR) of 30 individuals immediately after getting up in the morning. The participants in the study were all female athletes, between the ages of 18 and 29. In order to standardise the measurement, all HR values were determined after each participant had been sitting down for 5 minutes.

By examining column 2 in Table 8.3, it is difficult to arrive at any conclusions from the data. Importantly, it is also difficult to determine whether any erroneous values have been entered into the column. For example, to find the middle value would be a difficult task, due to the organisation of the data in the column. Likewise, the minimum and maximum values could not be obtained at a glance. However, when the data are organised through a ranking system, which in the case of column 4 is in ascending order, it becomes easier to locate the middle, minimum and maximum values at a glance. From column 4 it is clear that the middle value is between the 15th and 16th ranked value, since the column contains 30 observations. The middle value is therefore 58.5 beats per minute (beats·min^{-1}). Since heart rate is measured in whole beats per minute, this value would normally be rounded up to 59 beats·min^{-1}. The minimum and maximum values are located easily from the top and bottom values, respectively, in column 4, i.e. 43 and 73 beats·min^{-1}. The middle value is known as the **median**, and provides a measure of the central tendency of the data. By taking the minimum from the maximum value (i.e. 73 − 43 = 30 beats·min^{-1}), a measure of the variation of the data is provided, which is known as the **range**.

Continuing to think about the number of observations, questions also arise about the location of proportions of the data about the median value. For example, it might be useful to know how far from the median the highest and lowest quarters (or quartiles, Q) of the data reside. This value, known as the quartile range, is useful in further examining the spread of the data. So for the data in Table 8.3, the lower quartile (or 25% of the observations) is determined as the observations ranked 1 to 7.5 in column 4. To determine the value corresponding to the 7.5 observation, the 7th and 8th observations are added, and the result then divided by 2, i.e. (52 + 55)/2 = 53.5 beats·min^{-1}. Again, the figure must be rounded up to the nearest whole number (i.e. 54 beats·min^{-1}). Since the median value is 59 beats·min^{-1}, a quarter of the data reside 5 beats·min^{-1} below the median. Likewise, the upper quartile of the data can also be easily determined from column 4, as residing 5 beats·min^{-1} above the median. This is because the 22.5 observation (i.e. highest 25% of observations) is above 64 beats·min^{-1}. The inter-quartile range, which provides an insight into the spread of the data, is calculated by subtracting the lower quartile cut-off value from the upper quartile cut-off value, i.e. $64 - 54 = 10$ beats·min^{-1}. Another way of looking at this situation is that 50% of the observations are within a range of 10 beats·min^{-1} of the median.

From column 4, it is also possible to determine the most frequently observed heart rate. This value is known as the **mode**, and provides a further measure of the central tendency of the data. From column 4, it is apparent that 59 beats·min^{-1} was recorded on four occasions, which was more frequent than any other value. It is often easier to find the mode if a **histogram** of the data is produced. A histogram of the data from Table 8.3 is shown in Figure 8.2.

The histogram also provides us with a graphical representation of the distribution of the data (frequency distribution). It reinforces the information presented in tabular form. For example, it is clearly shown that a large proportion (i.e. at least 50%) of the observations are between 54 and 64 beats·min^{-1}.

All of the measures of spread and central tendency of the data presented so far are suitable

Table 8.3 Calculation of descriptive statistics based upon resting heart rate (HR, beats·min^{-1}) from 30 female participants

Participant	HR	Rank	Ranked HR	Descriptive statistics
1	59	1	43	Median: 59
2	68	2	46	Minimum: 43
3	55	3	47	Maximum: 73
4	57	4	49	Range: 30
5	51	5	50	Lower quartile: 54
6	59	6	51	Upper quartile: 64
7	55	7	52	Mode: 58
8	71	8	55	Mean: 58
9	61	9	55	Standard deviation: 7
10	58	10	55	
11	46	11	56	
12	59	12	57	
13	49	13	58	
14	62	14	58	
15	73	15	58	
16	64	16	59	
17	58	17	59	
18	43	18	59	
19	55	19	59	
20	59	20	61	
21	67	21	61	
22	61	22	62	
23	64	23	64	
24	52	24	64	
25	50	25	64	
26	47	26	67	
27	58	27	68	
28	72	28	71	
29	64	29	72	
30	56	30	73	

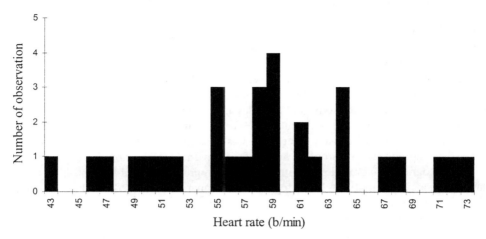

Figure 8.2 Histogram showing the frequency of observation of the range of heart rates

for description of ratio, interval and ordinal data. The mode is also suitable for a description of nominal data. With interval and ratio data, further tools are available to describe the centrality and spread of the data.

If all the data are added together and then divided by the number of observations, the **mean** average is determined (see Equation 8.1):

$$\text{Mean average } (\bar{x}) = \frac{\sum (x_1 + x_2 + ...)}{n} \tag{8.1}$$

where $\bar{x}$ is the mean average, Σ is the sum of observations, x is an individual observation (where x_1 is observation 1, etc.), and n is the number of observations.

The mean provides a further measure of the centrality of interval or ratio data. For the data presented, the mean is determined by adding the heart rate values for all participants, and then dividing by the number of participants, giving a value of 58 beats·min^{-1}. The advantage of the mean as a measure of central tendency is that all the observed values are taken into account in its determination, unlike the other measures. The mean therefore provides a better average of the data in general, and is the preferred measure of central tendency.

At this point, it is worth comparing the three measures of central tendency determined for the heart rate data. From Table 8.3 it is clear that the median, mode and mean provide similar values of 59, 58 and 58 beats·min^{-1} respectively. Comparisons of these measures of central tendency provide a simple check to determine the distribution of the data. A pertinent question is, why is the distribution of the data important? Often a prerequisite for inferential data analysis is that the data have a **normal distribution** (see Figure 8.4). A distribution is considered to be 'normal' when the data are distributed evenly about a central point, as usually determined through visual inspection of a histogram. A further check is to ascertain whether the central points as determined by the three methods of calculating central tendency agree. If the mean, median and mode all produce similar values, then the data are regarded as being normally distributed.

With interval and ratio data, it is possible to ascertain additional information regarding the spread of the data about the central value. The **mean absolute deviation** (MAD) is one such measure. To determine the MAD, the difference between each individual observation and the mean is determined, and then these differences are expressed as positive values (see Table 8.4). The sum of the differences is then determined. (If the differences were expressed as positive and negative values, rather than just positive values, the differences would cancel each other out when summed together.) Finally, the sum of the positive difference values is divided by the number of observations (see Equation 8.2).

$$\text{Mean absolute deviation (MAD)} = \frac{\sum \left(|\bar{x} - x_i| \right)}{n} \tag{8.2}$$

where x_i is each individual observation. Table 8.4 shows that, for the chosen example, the mean absolute deviation is 6 beats·min^{-1}. The higher the MAD, the greater the spread of the data.

The **variance** is another measure of the spread of the data, and is determined by finding the difference between the mean and each individual observation. Each difference value is then squared. The average of all the squared difference values is then determined by adding together all the squared differences and dividing by the number of observations (see Equation 8.3).

$$\text{Variance} = \frac{\sum \left(\bar{x} - x_i \right)^2}{n} \tag{8.3}$$

Table 8.4 shows that for the chosen example the variance is 56. A value known as the **standard deviation** (SD) may then be determined by taking the square root of the variance; thus in this example the SD is 7 beats·min^{-1} (see Equation 8.4). Both the variance and the SD value are used to describe the degree of spread of the data.

$$\text{Standard deviation (SD)} = \sqrt{\frac{\sum \left(\bar{x} - x_i \right)^2}{n}} \tag{8.4}$$

In order to understand more fully the significance of the SD, it is sometimes useful to represent graphically the tabulated information. Figure 8.3 shows lines representing one SD from the mean and two SDs from the mean in relation to the mean average of the individual observations, and the individual observations themselves. The lines on the graph illustrating one SD from the mean provide markers between which about 67% of observations from the sample lie. The lines on the graph representing two SDs from the mean provide markers between which about 95% of observations from the sample lie. The lines on the graph representing one and two SDs are referred to as the 67% and 95% confidence intervals respectively. The SD values are normally written in text by initially stating the mean average (i.e. 56 beats·min^{-1} in this example) plus or minus the SD (i.e. $\bar{x} \pm \text{SD}$ is 56 ± 7 beats·min^{-1}).

The SD is sometimes expressed in relation to the mean, resulting in a value known as the **coefficient of variation** (CV). To calculate the CV, the SD is divided by the mean, and then

Table 8.4 Calculation of variance, standard deviation and mean absolute deviation based upon resting heart rate (beats·min⁻¹) from 30 female participants

Heart rate	$\bar{x}$	$\bar{x} - x_i$	$(\bar{x} - x_i)^2$	$\sqrt{(\bar{x} - x_i)^2}$	Descriptive statistics
59	58	−1	1	1	Variance: 56
68	58	−10	100	10	Standard deviation: 7
55	58	3	9	3	Absolute deviation: 173
57	58	1	1	1	Mean absolute deviation: 6
51	58	7	49	7	
59	58	−1	1	1	
55	58	3	9	3	
71	58	−13	169	13	
61	58	−3	9	3	
58	58	0	0	0	
46	58	12	144	12	
59	58	−1	1	1	
49	58	9	81	9	
62	58	−4	16	4	
73	58	−15	225	15	
64	58	−6	36	6	
58	58	0	0	0	
43	58	15	225	15	
55	58	3	9	3	
59	58	−1	1	1	
67	58	−9	81	9	
61	58	−3	9	3	
64	58	−6	36	6	
52	58	6	36	6	
50	58	8	64	8	
47	58	11	121	11	
58	58	0	0	0	
72	58	−14	196	14	
64	58	−6	36	6	
56	58	2	4	2	

the answer is multiplied by 100. By multiplying by 100, the answer is expressed in the form of a percentage (%) (see Equation 8.5).

$$\text{Coefficient of variation (CV) (\%)} = \frac{\sqrt{\sum (\bar{x} - x_i)^2 / n}}{\bar{x}} \cdot 100 \qquad (8.5)$$

The CV is a standardised measure of the variation, and is therefore very useful when comparing the variation between two sets of data. In the heart rate data under consideration, the coefficient of variation is 12.5% (i.e. (7/56) × 100). If the temperature of the female subjects in our sample was also measured, in addition to heart rate, it would be possible to compare the amount of standardised variation in both measures.

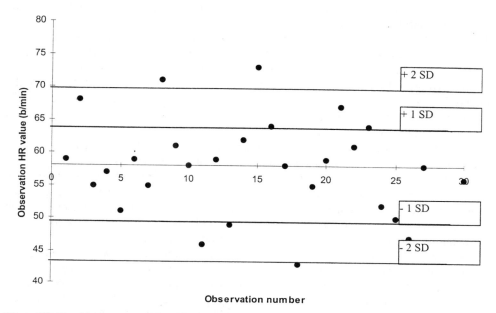

Figure 8.3 Graphical representation of standard deviation in relation to individual heart rate observations

At this point it should be stressed that although the SD and MAD both provide a measure of spread for the sample under consideration, the SD is known to be a more reliable representation of the spread of data in the population from which the sample was taken. It is for this reason that the SD is more widely used to illustrate the spread of data within a sample.

The importance of the relationship between a sample and the population from which it was taken becomes more apparent as we move on to drawing inferences from findings. When we need to infer something from the data, the population characteristics are always used as the reference point. Although a sample of only 30 participants was taken from the population in the earlier example, it is known that with that number of participants, providing representative sampling has taken place, the descriptive characteristics of the sample will be similar to those of the population.

Inferences from findings

In the main, two things are done when we draw inferences from data: either relationships or differences between data sets are determined. However, the **population** from which the **population** was derived is always taken as the reference point, and findings are related to the population. Let us now consider how particular questions are addressed.

To check whether a single observation belongs to a population, information is required about the mean and SD of the population. It is then necessary to examine the difference between the population mean and the observation, in relation to the population SD. It has already been intimated that if an observation is less than the value of one SD from the mean of the population, it can be stated with confidence that the observation is not from another population. If the observation is less than two SDs from the population mean it can still be stated with confidence that the observation is not from another population. However, if the observation falls outside the value of two SDs away from the population mean, it is likely that the observed value is not from the population. In fact, it can be stated with 95% confidence that the observed value is from another population. If the observation falls outside the value of three SDs away from the population mean, it can be stated with 99% confidence that the observed value is not from the population. It is sometimes easier to understand the concept of **confidence intervals** if a graphical representation of the population under consideration is provided (Figure 8.4).

Figure 8.4 shows what is known as a standard normal curve. This is the curve produced when, for standardised data, the standard deviation number is plotted on the x-axis in relation to the position of the mean. Earlier, it was mentioned that to ascertain whether an observation is from a population the difference between the value of the observation and the population mean should be determined in relation to the population SD. For example, if the observation value falls outside two standard deviations (± 2 SD in Figure 8.4), it is with 95% confidence that it can be stated that the observed value is not from the population.

In the area constrained below the curve, 67% of the observations reside between ± 1 SD, 95% between ± 2 SD, and 99% between ± 3 SD. Although 67%, 95% and 99% have been the only confidence intervals mentioned so far, it is possible to find a specific confidence level for any observation. Such confidence levels are determined by relating the standardised observed value to the standardised normal curve (Figure 8.4). Tables of values have been produced for this purpose. The values obtained from such tables simply relate to the proportion of the curve below or above the standardised value respectively.

It is sometimes possible to specify that if the observed value is not from the population, it is from a population with a greater mean average, or from one with a smaller mean average. If this is the case, only half of the curve is relevant for comparative purposes. For example,

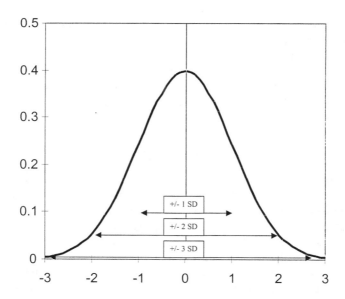

Figure 8.4 Graphical representation of a standard normal curve

if it was stated with confidence that the observed value was either from the population under consideration or from a population with a greater mean average, only the right hand side of the curve would be of relevance. In that case, 5% of the population observations are to the right of the 95% confidence interval on the right hand side of the curve. By contrast, when the whole curve was considered, because a directional effect could not be specified, only 2.5% of the population observed values would be contained in the same area. Tests to evaluate whether an observed value is from a different population are known as one tailed if the direction of the alternative population can be specified, and two tailed if the direction of the alternative population cannot be specified.

Although so far the value of a single observation in relation to a known population has been considered, a sample in relation to a population could also be considered, to establish whether the sample is from the population. Consideration could also be given to whether two samples come from the same population. The only difference when a sample is considered, rather than an individual observation, is that a population made up of several samples is discussed, rather than several individual observations. Each sample is assumed to have a similar SD, but a different mean value. In this case, rather than the population SD, a value known as the **standard error of the mean** (SEM) is used. So rather than a standard deviation of observed values within a population, a standard error of sample means is considered. In the same way that the difference between an individual observed value was compared with the population mean relative to the SD, the sample mean is compared with the population mean in relation to the SEM.

As well as differences between data sets, relationships between data sets often require consideration. For example, is there a relationship between the number of years spent in training, and the performance in a particular athletic event? Questions regarding performance and its relationship with another variable (i.e. training years in this example) are frequently asked in the study of sport and exercise. Although a non-linear relationship often exists

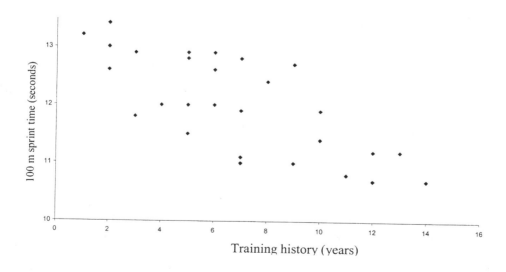

Figure 8.5 Scatter plot of years of training against 100 m sprint time

Table 8.5 Calculation of derived values based upon training duration (years) and performance times (seconds) of male 100 m sprinters

Years of training	$\bar{x}$	$\bar{x} - x_i$	$(\bar{x} - x_i)^2$	Performance time (s)	$\bar{y}$	$\bar{y} - y_i$	$(\bar{y} - y_i)^2$	$x_i \cdot y_i$
2	7	-5	25	13.4	12.1	1.3	1.7	-6.5
10	7	3	9	11.9	12.1	-0.2	0.0	-0.6
12	7	5	25	10.7	12.1	-1.4	2.0	-7.0
6	7	-1	1	12.9	12.1	0.8	0.6	-0.8
4	7	-3	9	12.0	12.1	-0.1	0.0	0.3
1	7	-6	36	13.2	12.1	1.1	1.2	-6.6
8	7	1	1	12.4	12.1	0.3	0.1	0.3
9	7	2	4	12.7	12.1	0.6	0.4	1.2
5	7	-2	4	12.0	12.1	-0.1	0.0	0.2
7	7	0	0	12.8	12.1	0.7	0.5	0.0
3	7	-4	16	13.6	12.1	1.5	2.3	-6.0
2	7	-5	25	13.0	12.1	0.9	0.8	-4.5
1	7	-6	36	13.2	12.1	1.1	1.2	-6.6
14	7	7	49	10.7	12.1	-1.4	2.0	-9.8
12	7	5	25	11.2	12.1	-0.9	0.8	-4.5
3	7	-4	16	11.8	12.1	-0.3	0.1	1.2
7	7	0	0	11.9	12.1	-0.2	0.0	0.0
5	7	-2	4	11.5	12.1	-0.6	0.4	1.2
7	7	0	0	11.1	12.1	-1.0	1.0	0.0
9	7	2	4	11.0	12.1	-1.1	1.2	-2.2
6	7	-1	1	12.6	12.1	0.5	0.3	-0.5
5	7	-2	4	12.9	12.1	0.8	0.6	-1.6
11	7	4	16	10.8	12.1	-1.3	1.7	-5.2
13	7	6	36	11.2	12.1	-0.9	0.8	-5.4
10	7	3	9	11.4	12.1	-0.7	0.5	-2.1
5	7	-2	4	12.8	12.1	0.7	0.5	-1.4
3	7	-4	16	12.9	12.1	0.8	0.6	-3.2
6	7	-1	1	12.0	12.1	-0.1	0.0	0.1
7	7	0	0	11.0	12.1	-1.1	1.2	0.0
2	7	-5	25	12.6	12.1	0.5	0.3	-2.5
Mean 7			Sum 401	Mean 12.1			Sum 22.7	Sum -72.5

between two variables in the study of sport and exercise, only linear relationships will be considered here. Linear models are used to describe linear relationships. Often the first step in establishing whether a relationship exists between two variables is to produce a **scatter plot**, such as that shown in Figure 8.5. This was produced from the data in columns 1 and 5 in Table 8.5. Such visual presentation of data is often required in order to observe a relationship between two variables.

From the scatter plot (Figure 8.5) it is clear that a relationship does exist between the number of years spent training and the 100 m sprint time. In fact, the relationship is what is referred to as a negative one, with one variable getting smaller as another gets larger. To see whether a relationship is negative or positive, imagine a line through the middle of the data: a so-called line of best fit. The slope of that line indicates whether the data are negatively or positively related. In the example shown in Figure 8.5, performance time gets smaller as the number of years spent training gets larger. A negative relationship is always shown by a negative slope on a scatter plot.

As well as the direction of the relationship (i.e. positive or negative), it is often necessary to test the strength of the relationship. The strength of the relationship is indicated graphically by how close the data points are to the line of best fit. The strength of the relationship can also be quantified, in terms of what is referred to as a **correlation coefficient** (r). In Table 8.5, all of the columns are used in the calculation of r. The columns show: the observed values (columns 1 and 5 respectively), the mean average of the observed values (2 and 6 respectively), the differences between the mean average and the observed values (3 and 7 respectively), the square of the differences (4 and 8 respectively), and the product of the observed values (column 9). The r-value is calculated by first determining the sum of the products of the observed values, and then dividing this value by the square root of the product of the sum of squares for both variables (see Equation 8.6).

$$\text{Correlation coefficient } (r) = \frac{SP}{\sqrt{SS_x SS_y}}$$

$$\text{where} \quad SP = \sum xy - \frac{\left(\sum x\right)\left(\sum y\right)}{n}$$

$$SS_x = \sum x^2 - \frac{\left(\sum x\right)^2}{n} \tag{8.6}$$

$$SS_y = \sum y^2 - \frac{\left(\sum y\right)^2}{n}$$

where SP is the sum of products, SS_x is the sum of squares for the x-variable, and SS_y is the sum of squares for the y-variable.

The r-value for the data in the present example is –0.76, indicating a negative relationship between number of years of training and performance in the 100 m sprint. The strength of the relationship is indicated by how close the r-value is to +1 or –1. In this example, –0.76 is quite close to –1, indicating a strong negative relationship.

In addition to inferences about differences between variables, inferences can also be drawn about the strength of a relationship between variables. The confidence in such inferences can also be determined. The correlation coefficient (r) from a sample can be compared with a population normal distribution of r-values to establish whether the r-value is derived from the population or not. Again, the confidence level with which the claim is made can take on a range of values. So it is possible to ascertain both the strength of a relationship, and whether the relationship was from a particular population. It is important to note at this point that the strength of a relationship and the significance of the relationship provide no insight into whether the relationship between the variables is causal. A causal relationship is one in which the change in one variable explains the change in the other variable.

Choosing an appropriate statistical test

In addition to learning about the basic concepts underlying interval and ratio data analysis, it is also necessary to have a knowledge of the range of appropriate data analysis tools available. With an understanding of the concepts underlying statistical treatment of data, an informed decision can be made about the most appropriate test for a particular purpose. However, it is not essential to understand how to perform each statistical test by hand, as such tests are usually performed with the help of purpose-designed statistical packages (see Chapter 9 for details).

A range of tests are available for determination of differences between samples. The tests detailed below are suitable for a variety of situations where data are measured on an interval or ratio scale.

The independent t-test is used to determine differences between two samples when the samples do not include the same participants. It is therefore not possible to pair the data for analysis, so the descriptive characteristics of the complete samples are used in the analysis. Whether a one-tailed or a two-tailed test is used will depend on the hypothesis being tested. Earlier in the chapter it was mentioned that if the direction of change is stated in the hypothesis, a one-tailed test should be used. At the same level of significance, the choice of a one-tailed

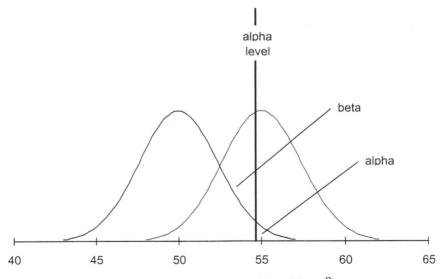

Figure 8.6 Area of normal distributions defined as alpha (α) and beta (β)

test will increase the chance of finding a difference if one exists. A one-tailed test is, therefore, considered a more powerful test (see 'Statistical errors and power' section below).

The related t-test is used to determine differences between two samples when the samples include the same participants. Such an experimental design is referred to as being repeated measures. For analysis, the data are treated as paired data, since the same participants appear in both samples. The related t-test is sometimes referred to as a paired t-test. A related t-test is considered to be more powerful than an independent t-test. Again, with this type of test, a one- or two-tailed test should be chosen, depending upon the wording of the hypotheses.

The independent measures analysis of variance is used to determine differences between two or more samples, when the samples include different participants. Again, it is not possible to pair the data in this situation, so whole sample characteristics are used in the analysis. Since with this type of test two or more samples are compared, a further test is necessary to determine the location of any significant differences, if there are any. Such tests are known as post-hoc tests, and are named after their inventor, for example Tukey, Scheffé, Newman–Keuls.

The repeated measures analysis of variance is used to determine differences between two or more samples when the samples include the same participants, i.e. paired data. As with a related t-test, a repeated measures analysis of variance is more powerful than an independent measures analysis of variance. Again, a post-hoc test must be used to locate any differences demonstrated with the analysis of variance.

It is important to note that the tests mentioned above should be used only when the data conform to certain standards. These prerequisites include that the data should be of interval or ratio type, and normally distributed. If the data are not normally distributed, inferences may still be drawn, by employing other tests which are more usually used with ordinal data. In this situation, the independent t-test can be replaced by the Mann–Whitney test, the related t-test by the Wilcoxon test, the independent measures analysis of variance by the Kruskal–Wallis test, and the repeated measures analysis of variance by the Friedman test.

Statistical errors and power

Statistical errors occur when the incorrect hypothesis is accepted/rejected based on the available data. For example, if the alternative hypothesis is accepted, but the null hypothesis should have been accepted, this is known as a type I **statistical error**. In other words, if a difference is claimed between two sets of data when there is no difference, a type I statistical error has been made. Alternatively, if the null hypothesis is accepted, but the alternative hypothesis should have been accepted, this is known as a type II statistical error. In other words, if it is claimed that no difference exists between data sets when a difference does exist, a type II statistical error has been made.

The question remains, what factors lead to these inappropriate conclusions? The answer to this question is best viewed graphically. From Figure 8.6 it can be seen that the chosen confidence level (alpha level) defines the alpha (α) area under the left curve. The beta (β) area is then the area to the left of the α level constrained by the right curve. The β area is the proportion of the right curve that passes the pre-defined α level.

By relating the α and β areas to hypothesis testing it is possible to see how statistical errors are made, and more importantly how statistical errors may be prevented. If the chosen α level is too relaxed, thereby resulting in a large α area, the chances of concluding that there is a difference, when no difference is actually present, are increased. So a relaxed α

level may result in a type I statistical error. Likewise, if the β area is too large, there is an increased chance of concluding that there is no difference, when a difference is actually present. A β level is never set prior to performing a statistical test, however, so this situation must be examined in a slightly different way.

Rather than specify that the β area must be as small as possible, the area to the right of the α level which is constrained under the right curve can be maintained as large as possible. This area is quantified as being the area under one whole curve minus β (i.e. $1 - \beta$), and is also referred to as the power of the test of differences. So the power of a test is defined as the area $1 - \beta$. Therefore, rather than talk about the β level, the power can be considered. The power of the test is what influences the chances of making a type II statistical error.

Whilst the size of the α level may be specified prior to performing a statistical test, it is not so simple to change the power $(1 - \beta)$. Three factors influence the power:

- the number of participants (n);
- the effect size (γ) of the treatment;
- the alpha (α) level.

The more participants in a study, the greater the power. This is because as the participant number is increased, the shape of the normal distribution curve changes, effectively becoming narrower. A narrower curve results from less variation about the mean, so less overlap is likely between curves.

The effect size (γ) is the difference between the means in relation to the variation in the data (Equation 8.7).

$$\text{Effect size } (\gamma) = (\bar{x}_1 - \bar{x}_2)/\text{SD}. \tag{8.7}$$

If the difference between the means is large in relation to the variation in the data, γ will be large, which will increase the power. Similarly, if the variation in the data is reduced, γ will be increased, thereby increasing the power. Both an increase in the difference between means, and a decrease in the variation about the means, lead to less overlap between the normal distribution curves.

The more relaxed the α level, the greater the power. However, by increasing power through relaxation of the α level, the chances of making a type I error are increased. By relaxing the α level, the overlap between the curves is not reduced. This would, therefore, be the least desirable way to increase the power of the test. Thus the preferred methods are increasing participant numbers (n) and/or effect size (γ), without decreasing the α level. In this way, the chances of making both a type II and a type I error are minimised.

Although it is not necessary to know the intricacies of the calculation of statistical power, it is useful to have an idea of the factors which influence it. With this knowledge, it is possible to interpret results from studies appropriately. For example, a study that found no statistical difference between two groups, but had only five subjects, should be investigated further in terms of the power to detect a difference if one was indeed present. In this example, the power may be found to be low, in which case the alternative hypothesis cannot be rejected with confidence. Knowledge of statistical errors is also required when designing experiments. Often investigators will perform prospective power analysis to establish the required subject numbers.

Assessing reliability and validity

An investigator should ensure that measurements in studies of sport and exercise are both reliable and valid. Since measurements in laboratory work in sport and exercise are usually objective, inter-investigator reliability is usually good. Inter-investigator reliability will usually be good if direct measurements are taken from equipment with no requirement for interpretation, providing equipment is properly calibrated. Inter-investigator reliability may be examined by asking two or more investigators to make the same measurement under identical conditions. If the measurement or data collection technique is complicated, or requires skills developed through training, inter-investigator reliability may be poorer. It is, therefore, important to examine the method used to assess the reliability of the equipment and/or the data collection technique and/or the response of the participant.

A number of data collections should be made, using identical equipment and data collection techniques and with the participant maintaining a stable response. These data can then be assessed to determine the mean average and standard deviation, and based on these values the coefficient of variation may be determined (i.e. (SD/mean) $\times$ 100). The acceptable reliability will be based on the expected change in a study. For example, if it is expected that a 9% change will be observed, a coefficient of variation of 5% may be acceptable; however, if a 2% change is expected, a coefficient of variation of 0.5% may be required.

Data cannot be valid if they are not reliable. This is because it is impossible to make a valid measurement if the measurement cannot be shown to produce the same result time and time again under the same conditions. Normally, tests of validity compare one measurement result with another measurement result to determine if the two results agree. The extent to which the results agree determines the degree of validity of the measurement. Two methods for determining validity have already been covered in this chapter. One method requires testing for statistical differences between two measurement samples, one being a criterion measurement. If no difference is observed between the two samples, then the measurement being tested for validity can be termed valid in relation to the criterion measurement. The other method uses the relationship between measurements to determine validity. Again, one measurement is related to a criterion measurement. If the correlation between the two measurements is strong, and it is shown that the two samples are significantly similar, the tested measurement is said to be valid in relation to the criterion measurement. A problem with this method of assessment of validity is that the strength of the relationship is markedly influenced by the range in the data.

ACTION POINTS

1 Describe inter-investigator reliability, and explain how it may be improved.
2 What is the aim of scientific investigation?
3 Present an example of a null hypothesis and an alternative hypothesis.
4 What is the relationship between hypotheses and data analysis?
5 What are the two key points to consider when designing experiments?
6 What are the four types of data?
7 Which of the four types of data are categorical and which are continuous?
8 What are the two types of data analysis?
9 List three measures of central tendency.

10 List five measures of spread.
11 Which measure of central tendency takes account of all the observed values?
12 Which measure of central tendency is suitable for use with categorical data?
13 Which measures of spread are suitable for use with continuous data only?
14 What is the difference between the standard deviation and the standard error of the mean?
15 Which measure of central tendency and spread is used to determine the coefficient of variation?
16 How is a distribution of data tested for normality?
17 What statistic is used to compare values relative to their respective populations?
18 What is the difference between a one- and a two-tailed test?
19 How are hypotheses related to one- and two-tailed tests?
20 What is a confidence interval?
21 How are confidence intervals used in statistical tests?
22 How are confidence intervals related to the chance of making a type I error?
23 Show graphically how the power of a test is determined.
24 List three methods of determining the validity of a measurement.
25 Can a measurement be valid if it is not reliable?

Conclusion

Data analysis cannot be discussed in isolation from experimental design and hypothesis testing. Experiments should not be designed without giving considerable thought to the required data analysis. Additionally, hypotheses are tested through appropriate data analysis, and conclusions may be misleading if inappropriate data analysis is performed. Particular attention must be given to the type of data to be analysed, specifically whether the data are nominal, ordinal, interval or ratio level. Often data analysis is broadly divided into descriptive and inferential analysis. Descriptive analysis allows information to be gleaned from a set of data quickly. Inferential analysis allows claims to be made about the sample of data in relation to a population or another sample. Inferences about either the similarities or differences between samples, or the relationship between samples, are possible. Such claims are then used in order to accept or reject hypotheses with a degree of confidence. The confidence interval is set by the investigator prior to the study, and often a 95% confidence interval is selected. The chance of making a type I statistical error is then small (i.e. 5%). The chance of making a type II statistical error is assessed through the power of the statistical test, and it is normally advised to increase this figure to 80% (i.e. 20% chance of making a type II error). Descriptive data analysis may also be used to assess the reliability of measurements, whilst inferential data analysis may be used to assess the validity of measurements. The importance of a knowledge of data analysis for a student of exercise and sport cannot be over-emphasised.

KEY POINTS

- Scientific enquiry is based upon the formation of testable hypotheses.
- Hypotheses are tested through inferential data analysis.
- Data can be categorised as being of four types: nominal, ordinal, interval or ratio.

- Descriptive data analysis should present required information to an observer in an easily accessible form.
- Measures of central tendency and spread are fundamental to descriptive data analysis.
- Many inferential statistical tests are dependent upon normally distributed data.
- When a direction of change is hypothesised, a one-tailed test should be used.
- A type I statistical error is when a difference is accepted, although no difference actually exists.
- A type II statistical error is when no difference is accepted, although a difference actually exists.
- Measurements implemented in scientific investigations should be both reliable and valid.

Bibliography

Derzin, N.K. and Lincoln, Y.S. (1994) *Handbook of Qualitative Research*. London: Sage Publications Ltd.

Derzin, N.K. and Lincoln, Y.S. (1998) *Collecting and Interpreting Qualitative Materials*. London: Sage Publications Ltd.

Hinton, P.R. (1995) *Statistics Explained: A Guide for Social Science Students*. London: Routledge.

Howell, D.C. (1997) *Statistical Methods for Psychology*. Belmont: Wadsworth Publishing Co.

Kreider, R.B., Miller, G.W., Williams, M.H., Somma, C.T. and Nasser, T.A. (1990) Effects of phosphate loading on oxygen uptake, ventilatory anaerobic threshold, and run performance. *Medicine and Science in Sport and Exercise* 22(2): 250–256.

Patton, M.Q. (1980) *Qualitative Evaluation Methods*. Beverly Hills, CA: Sage Publications Ltd.

Silverman, D. (1993) *Interpreting Qualitative Research*. London: Sage Publications Ltd.

Vincent, W.J. (1995) *Statistics in Kinesiology*. Champaign, IL: Human Kinetics.

9

INFORMATION AND COMMUNICATION TECHNOLOGY (ICT)

AIMS OF THE CHAPTER

This chapter aims to provide an understanding of information and communication technology which may be applied to the study of exercise and sport. After reading this chapter you should be able to:

- illustrate the importance of information and communication technology (ICT) in exercise and sport;
- outline user friendly developments in ICT;
- demonstrate the ability of ICT to improve work in exercise and sport;
- identify the opportunities provided through a networked computer;
- provide examples of ICT use in exercise and sport.

Introduction

Information and communication technology (ICT) is the technology now available on personal computers which allows processing and storage of digital information, and communication between computers. The term ICT is now widely used in many areas, including academia, and has largely replaced the term information technology (IT). The replacement of the term IT is significant, since it is a statement that communication between computers (communication technology, CT) is fundamental to the way in which computers are used today. The day of the personal computer as an isolated piece of equipment has now passed.

The days when the study of sport and exercise and the use of ICT were two independent, unrelated tasks are also long gone. The use of ICT permeates all aspects of the study of exercise and sport, as is the case with most other academic subjects. Not only is the use of ICT necessary to study exercise and sport effectively, but employers of students of exercise and sport expect ICT skills. Skills in ICT are just as important as oral, written and numerical skills. In fact, all such transferable skills are inter-related, and oral, writing and numeracy tasks are often aided through the use of ICT.

In order to develop ICT skills, it is not appropriate simply to read a book on the subject. Rather, the best way to learn is through first-hand experience. The systems now used on computers are designed in a way that encourages experiential learning. They are designed so that the user is faced with a range of options, and rarely has to remember a command. Hopefully, this chapter will provide enough information to give an insight to those with no experience. It will also provide examples of the way in which different software can be applied to the study of exercise and sport.

The term ICT covers a broad range of issues, ranging from the specific details of a personal computer to the access to a very large source of information – the **Internet**. In this chapter some aspects of ICT that are most relevant to a student of exercise and sport will be covered. Information on the specifications of a personal computer will therefore be left to the computer experts, and instead attention will be concentrated on the application of **software**. The application of software extends to software that is used to send and receive email and surf the Internet. When a personal computer is referred to, in many cases it will be part of a **network**. A networked computer is simply one that is connected to a telephone line, along which digital information may pass. For this facility, the computer must have a device known as a **modem**, which converts computer signals into audio signals for subsequent transmission down a phone line. Many personal computers now have a modem fitted as standard.

The study of exercise and sport is certainly enhanced through the use of ICT. For example, scientific reports are normally produced with the use of a word processing package, and often include reference lists which have been produced using a database package, statistical results that have been produced using a statistical package, and graphs that have been produced using a graphics package. If the report is published in the scientific literature, an email address for the author will often be included for correspondence. If a student is working in the laboratory, data may be collected using equipment that is linked directly to a personal computer, or the student may collect data and then enter the results into the computer. If a student is working in the library, information about how the human body works may be gained from an interactive CD-ROM which provides real-life pictures of the physiological systems at work. The student may also search for abstracts or full papers on the Internet, since most academic journals are now published in a digital form. If information is required about sport and exercise organisations, the Internet can usually provide details. Throughout this chapter, it will be demonstrated how software may be put to some of these uses.

The personal computer

Most students entering higher education from school will be familiar with the concept of a personal computer. Fortunately the days when commands had to be remembered in order to execute a task have passed. Nowadays the personal computer will run **menu-driven software**, which simply means that a task is selected from a menu of tasks. New personal computers are capable of running menu-driven software, regardless of the type of computer.

The computer system

The two most common types of **computer system** installed on a personal computer are Windows, which is produced by Microsoft, and the Apple Macintosh system. Both systems have become very compatible, and documents produced using one system can now usually be read by the other system. All that such a system does is enable the computer to run particular software.

A modern system installed on to a personal computer is designed to make the organisation of work simple. A single piece of work that is saved on to either the hard **disk** (C drive) or floppy **disk** (A drive) is known as a **document**. So if a report was produced using a word processing package, that report would be known as a document. This is analogous to a printed hard copy of the report, which would also be referred to as a document. If the report was printed out, it might be decided to place it into a **file**, along with the other information related to the report. Using the computer system, it is possible to do exactly the same, i.e. file a document. You may also decide to file the file containing the report within a filing cabinet. Once again it is possible to carry out that same procedure on the computer system, by operating levels of files. Figure 9.1 illustrates some of the above principles, along with the type of information that may be viewed on a computer screen.

From Figure 9.1 it is clear that the icon, or symbol, for documents created with different types of software looks different. For example, compare the icon for an Excel document with that for a Word document. Simply by double clicking (clicking twice in rapid succession) on one of these icons with the left button on the **mouse** a document can be opened. To move a document click once on the document and drag it to its new location by holding the left

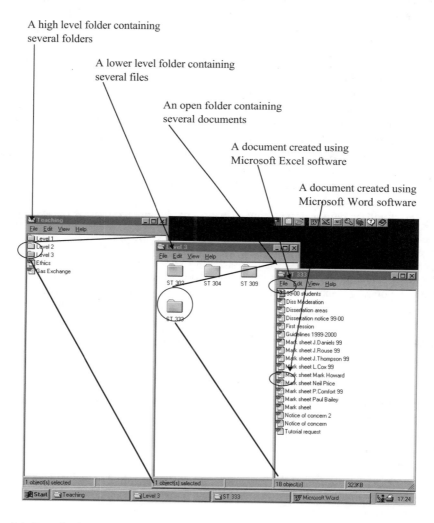

Figure 9.1 Organisation of documents within folders, and folders within folders. Microsoft Office® products, including Word® and Excel® © 1995 Microsoft Corporation. Used with permission (Microsoft Office, Word and Excel are trademarks of Microsoft)

button on the mouse down. It is always easier to try such things rather than imagine what needs to be done.

This section has provided a brief insight into organising the 'desktop' of a computer through the management of files, and documents within the files. Once a document has been opened, it is then necessary to consider the specific software used to create the document.

Software

In computer terminology, an item of software is often referred to as an **application**, and this term will be used frequently in this chapter. Most software produced now runs on the two systems mentioned above (Windows or Apple Macintosh), which have a similar appearance. Once working within an application the rules for executing a task are similar. Most software includes menus that are selected to perform a task. An example of a menu is given in Figure 9.2. To execute a task, the left mouse button is depressed when the mouse marker is over the menu bar. By keeping the mouse button depressed, the appropriate task can be selected. As the mouse button is released, the task is performed.

Various categories of software are available, including word processors, spreadsheets, databases, presentation packages, data analysis packages and graphics packages. Usually some limited data analysis and graphical functions can be performed within a spreadsheet package. Importantly, it is relatively simple to insert information from one type of package into another. For example, if working within a spreadsheet, it is possible to copy the information and paste it into a word processed document or a presentation package. Some

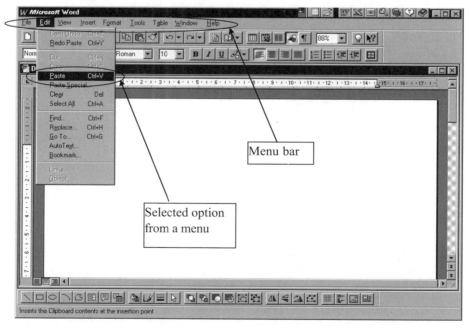

Figure 9.2 An example of a selection from a menu within an application. Microsoft Word® © 1995 Microsoft Corporation. Used with permission (Microsoft Word is a trademark of Microsoft)

information on specific software packages will now be provided, giving examples of their use in the study of exercise and sport.

Word processors

When there is a requirement to produce a report, an essay, a letter, or any other form of written document, word processing software is used. By using such an application, text can be organised into the most appropriate form. However, it is too simplistic to think that a word processing application is simply a computer-based typewriter. A word processor can allow 'better' or clearer writing, since it is possible to keep revisiting the work to make improvements. For example, it is very easy to cut and paste chunks of text. The cut and paste functions in a word processing application perform exactly the same tasks as if cutting and pasting text on a piece of paper. Text can also be selected in order to alter the format. For example, it may be desirable to make a word bold, put some text in *italics*, CAPITALISE some text, or increase the SIZE of text. The format of a paragraph may also be changed. For example, each line in a paragraph can be aligned only on the left side (what is known as left justified), or aligned on both sides (fully justified). Figure 9.3 illustrates a selection of options available under the Format menu in a word processing application.

The ability to move text around and to format the text often results in enhanced presentation, aiding the reader. Other useful tools now offered by most word processing applications allow spelling checks and grammar checks. Often these not only draw attention to the incorrect or suspect word or phrase, but also suggest a possible correct replacement.

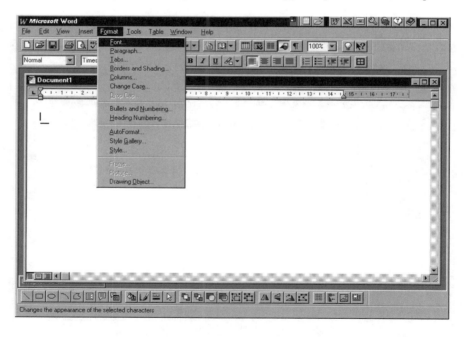

Figure 9.3 The list of options available under the Format menu in a word processing application. Microsoft Word® © 1995 Microsoft Corporation. Used with permission (Microsoft Word is a trademark of Microsoft)

A large part of science in the study of sport and exercise involves writing reports based on laboratory and field work (see Chapter 10). Word processing applications are essential for this type of work.

Spreadsheets

When dealing with numerical data, it is usual to enter the data into a spreadsheet. As well as allowing storage of the data, a spreadsheet enables organisation and manipulation of them. A spreadsheet is simply formed from rows and columns, which together form a series of cells. Each row has a reference and each column has a reference. Columns usually have a letter as a reference (i.e. A, B, C, etc.), and rows usually have a number (i.e. 1, 2, 3, etc.). Each cell consequently has its own unique reference (e.g. A1).

A series of data is normally entered into a column, and often the column is given a name. For example, in column A in Figure 9.4 the title is 'Participant number' and the participants are entered into the column in numerical order. The second column, column B, is titled 'Age', and the age of each participant is entered. Suppose it was then desired to find the mean average age of the participants. Such data analysis is performed by selecting the data analysis option from the Tools menu. A variety of data analysis options are available, many of which have been covered in Chapter 8.

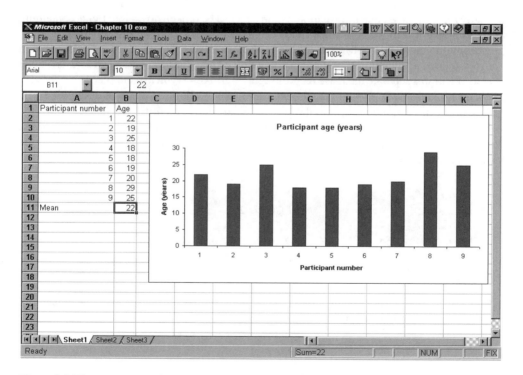

Figure 9.4 The appearance of a typical spreadsheet. The histogram is a graphical representation of the data shown in columns A and B of the spreadsheet. Microsoft Excel® © 1995 Microsoft Corporation. Used with permission (Microsoft Excel is a trademark of Microsoft)

In addition to sorting and analysing data, it is also possible to create graphical representations of them. Details of different types of graphs are given in Chapter 10, so only a single example will be provided here. In Figure 9.4, the data from the two columns were presented as a histogram by selecting the chart wizard icon. Although a histogram was produced in the example, it is also possible to create pie charts, scatter plots, line graphs, and numerous variations of these.

Once a graph has been produced using a spreadsheet, the graph can be copied and pasted into a word processing application. This facility is very useful, especially when preparing reports.

Databases

When dealing with information that would normally be organised in a filing system, a database is often the appropriate software. A **database** is simply a filing system, and each file is known as a record. Records can be sorted as a filing cabinet might be sorted, for example according to the number of the record. If the record does not contain a number, the records might be sorted alphabetically or in some other way.

Within each record, information is contained within fields. A field is equivalent to sections of a file that contain standard types of information. For example, if a database is used for storing and sorting references, one field may be used for the author's name, another may be used for the title of the paper, and another for the details of the journal (see Figure 9.5). It is normally recommended that information within a record be broken down into fields containing small amounts of information, since fields can always be merged but not separated.

An obvious application for a database is the organising of references. Often through study of an area a student will build up a collection of many references. Each reference can be placed within a separate database record; Figure 9.5 gives a typical example. The reference has been broken down into its constituent parts, which allows it to be re-assembled in a variety of forms (e.g. to conform to the style of a particular scientific journal).

For example, suppose a student is required to display a list of references according to the Harvard style (see Chapter 10). The fields from the record shown in Figure 9.5 may be easily assembled into the Harvard form:

RECORD 28

FIELD 1   Rowbottom D G Keast D Green S Kakulas B Morton AR

FIELD 2   1998

FIELD 3   The case history of an elite ultra-endurance cyclist who developed chronic fatigue syndrome

FIELD 4   Medicine Science Sport Exercise

FIELD 5   30(9): 1345–1348

Figure 9.5 An example of a database record for a reference

Rowbottom, D.G., Keast, D., Green, S., Kakulas, B. and Morton, A.R. (1998) The case history of an elite ultra-endurance cyclist who developed chronic fatigue syndrome. *Medicine Science Sport Exercise* 30(9): 1345–1348.

Once the desired records have been selected, it is possible to form a report (see Figure 9.6). A report is simply a way of organising selected records for a further purpose. The report can then be saved as part of a word processed document, or printed out. Once again, this facility is useful when writing scientific reports.

Presentations

In the area of sport and exercise, as with most other academic disciplines, oral presentations are an important way of disseminating information. Oral presentations are often supported by visual material. The utilisation of several types of information, including text, numbers, images, video and sound, is often referred to as **multimedia**. As the term suggests, multimedia is the use of several different mediums for getting information across to the observer/learner. Technology is now available to allow lecturers to pre-record a complete lecture, with video footage of the lecturer talking interspersed with other visual aids. The days of a lecture with a lecturer present may well be numbered!

It is often a requirement of an undergraduate course in exercise and sport that students present material to a group of students or staff. In addition to being a useful way of disseminating information, the ability to present is an important transferable skill. In many cases the visual aids are as important as the oral component of a presentation, and can enhance a presentation dramatically if they are well considered. Presentation packages to

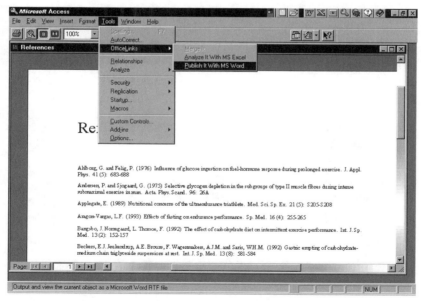

Figure 9.6 An example of a report produced using information from database records. Microsoft Access® © 1995 Microsoft Corporation. Used with permission (Microsoft Access is a trademark of Microsoft)

facilitate production of clear visual aids are available. Such packages enable production of overhead projector (OHP) transparencies, handouts, slides, notes pages for the presenter, and slide shows for projection directly from a personal computer. Whilst it is possible to produce visual aids without using a presentation package, such packages make preparation of presentation material considerably easier. Figure 9.7 shows an example of a slide for a presentation, and Figure 9.8 shows a handout, which is automatically created from several slides.

Many presentations are now made directly from a personal computer (i.e. a slide show), which has several benefits. The greatest benefit is the ability to include sound and video footage within the presentation, all from the computer. Additional benefits are the lack of requirement for video recorders and printing and photocopying facilities. Many academic departments now have facilities to project directly from a personal computer. Multimedia presentations are therefore easier, since video recorders and colour printing are not necessary. The common availability of scanners makes presentation directly from a computer even easier, since figures and tables can be scanned on to the hard disk of the computer from relevant papers and books. Slide shows also include an option to run themselves automatically, with prior timings entered for the display of each slide. If the presenter is not brave enough to use this facility, the slides can be progressed manually!

Data analysis

Although spreadsheet applications often include data analysis functions, these functions are usually limited. For a greater variety of data analysis functions, it is usually necessary to use

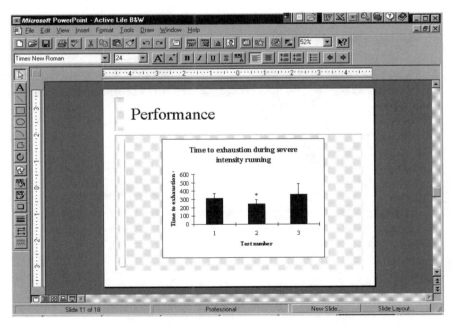

Figure 9.7 An example of a slide produced using a presentation package. Microsoft PowerPoint® © 1995 Microsoft Corporation. Used with permission (Microsoft PowerPoint is a trademark of Microsoft)

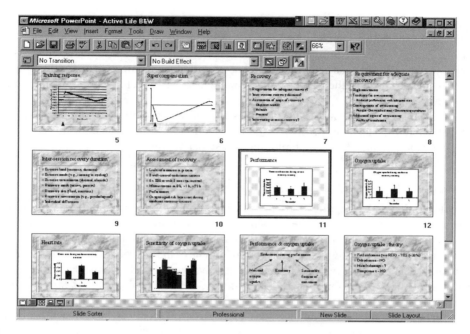

Figure 9.8 An example of a handout of slides created using a presentation package. Microsoft PowerPoint® © 1995 Microsoft Corporation. Used with permission (Microsoft PowerPoint is a trademark of Microsoft)

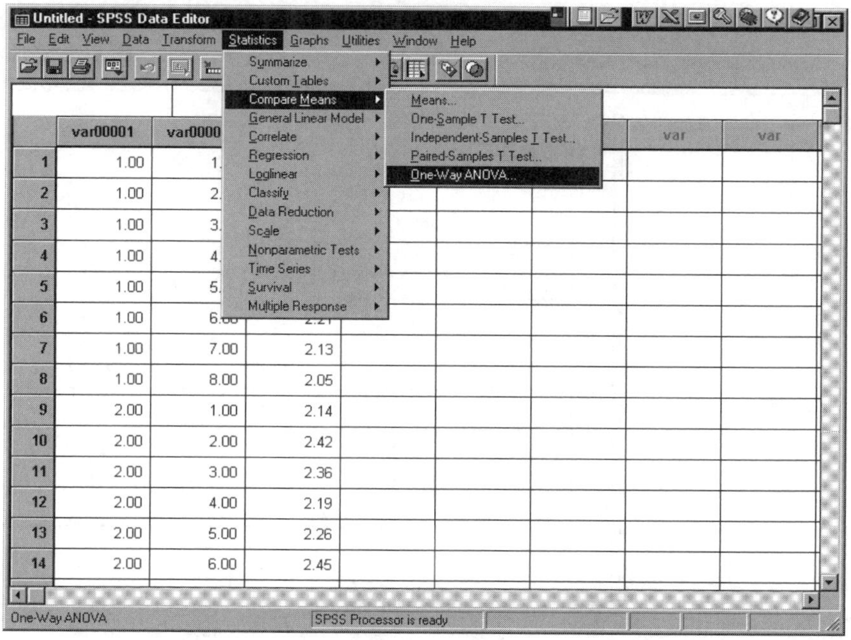

Figure 9.9 Parametric statistical tests available in a data analysis application. SPSS® © 1998 SPSS Incorporated. Used with permission (SPSS is a trademark of SPSS)

a specialised **data analysis application**. Figure 9.9 shows an example of a data sheet within a dedicated data analysis application.

Data analysis software normally includes a spreadsheet, where the data for analysis are entered (often referred to as a data sheet or data editor), and an output sheet for the result of any data analysis (often called a report sheet or output sheet). Figure 9.10 shows an example of an output sheet.

Data analysis software also often includes limited graph plotting capabilities. As was seen in Chapter 8, data analysis involves the production of graphs, so applications are now available which include both graph plotting and statistical functions.

Graphics

Although some graphics applications are now incorporated into data analysis applications, some specialised graphics software remains separate. For much undergraduate work in the area of exercise and sport, the graphics functions within standard spreadsheet applications provide adequate options. Figure 9.11 shows an example of the type of graph which specialised applications can produce. In this example, a curve has been fitted to some data collected as a participant starts exercising. A facility to fit various types of curves to data is useful in laboratory-based work.

The need for detailed and accurate graphs, particularly in laboratory-based areas of exercise and sport study, may sometimes require the use of specialised applications. Although students

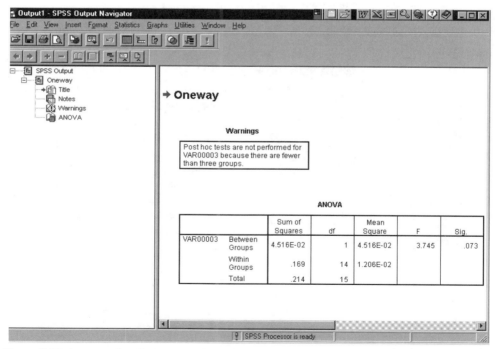

Figure 9.10 An example of an output sheet of a data analysis application. SPSS® © 1998 SPSS Incorporated. Used with permission (SPSS is a trademark of SPSS)

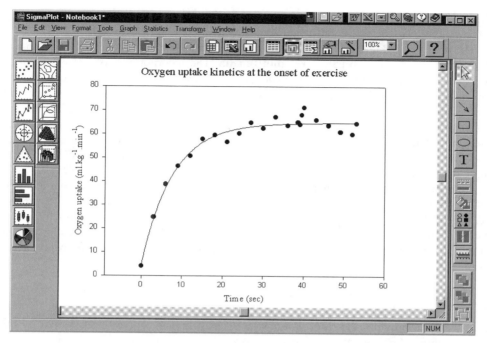

Figure 9.11 A graph produced using a specialised graph plotting package. SigmaPlot® © 1998 SPSS Incorporated. Used with permission (SigmaPlot is a trademark of SPSS)

in the early stages of their study will not be penalised for relatively simple graphs, more detailed graphs are often expected of final year students, particularly for their dissertation.

The Internet

The ability to process, store and retrieve digital information, and to pass digital information down telephone lines, has enabled the growth of communication technology, particularly the Internet. It is unlikely that communication systems will ever be the same following the introduction of this technology. Not only does the Internet enable messages to be sent from one computer to another (i.e. **electronic mail**, email), but it also enables access to an extremely large source of information, the **World Wide Web** (WWW). The only requirement for a networked computer to access information on the WWW is that it has a **web browser** installed. A web browser simply allows the computer to read files in Hypertext Mark-up Language (HTML). HTML is the common language used to produce most files on the WWW.

World Wide Web

The WWW was created in the late 1980s in a slightly different form to that which is available today. It was developed by European scientists who were trying to increase research collaboration. The predecessor of the WWW was called Gopherspace, and was text-only.

The standard tool for searching Gopherspace is known as Veronica. Some academic information in Gopherspace still remains to be converted to HTML form.

The use of HTML has given the WWW much wider appeal, since text may be combined with other images in web pages, i.e. **multimedia**. The incorporation of links (known as hyperlinks) to other similarly formatted documents (web pages) enables movement around information sources (surfing). A particular advantage of hyperlinks is the opportunity to arrange information in a non-sequential way. Rather than the designer dictating the order of information, the learner/observer can progress in their desired order. It is suspected that presentation of non-sequential information is an aid to learning. In a similar way that Windows-based computer systems are more user friendly, the graphical WWW interface is more user friendly.

An enormous web of information has been created in digital form. To find the required information within this large web there are various search tools. Obviously if the precise address of the required information is known, the information can be accessed without the use of search tools. For example, web addresses for possible useful web sites have been provided throughout this book. The address for one useful web site is given below as an example:

http://www.theses.com/registered_users/quick.html

This web site provides the titles and abstracts of all higher degree theses stored by the British Library. For students investigating a particular area, this may be a useful resource. A web site address usually starts with 'http' (which stands for hypertext transfer protocol), followed by a colon and two forward slashes and then 'www'. The remaining information is then specific to the web site.

As indicated above, a variety of search tools for the WWW are available. If you are searching from the UK or Europe it is often better to avoid the search engines based in the USA, since a transatlantic bottleneck may be experienced, resulting in slow response times. A selection of search tools which are detailed on the Cheltenham and Gloucester College home page are given below; information about each search tool is provided at http://www.chelt.ac.uk/category/search.htm.

- AltaVista
- Yahoo!
- Infoseek Guide
- SavvySearch
- Inference Find!
- CUSI
- WebCrawler
- TRFN

Once in a search tool, a key word(s) is normally required for the search. Once the key word is entered, the search engine will find all information on the WWW which includes that word. It is then possible to narrow down the search by providing further key words.

The Internet, and particularly the WWW, is now used widely in higher education as part of learning support. In many institutions, the Internet is used to provide independent learning opportunities in addition to taught elements of modules within courses. Some tutors write their own web pages to support taught aspects of a module or course.

Email

Electronic mail (email) is fast becoming an essential tool in the workplace, not just in academia, but in the world generally. The benefits of email are many: a major one is that the time taken for the sent mail to arrive at a destination is very short (a few minutes at the most) no matter where in the world the mail is sent. It is therefore possible for communication in a written form to take place almost instantaneously between a person at a personal computer in the UK and a person at one in Australia. Another benefit is that the mail is sent for the cost of a local telephone call, and the call lasts only as long as it takes to send the mail from the computer (usually a few seconds). In addition, other files, e.g. containing text, spreadsheets, databases or presentations, can be attached to the email message proper: the only requirement is that the person receiving the email has the appropriate software on their computer to allow them to open the attached files.

Email software is very simple to use. As well as sending and receiving messages, such applications allow the storage of received email messages, sent mail and deleted messages (see Figure 9.12).

Many higher education institutions now expect students to register for an email account. In this way students can correspond easily with colleagues and friends, as well as with their tutors. Some courses now include email discussion groups, where students can ask questions,

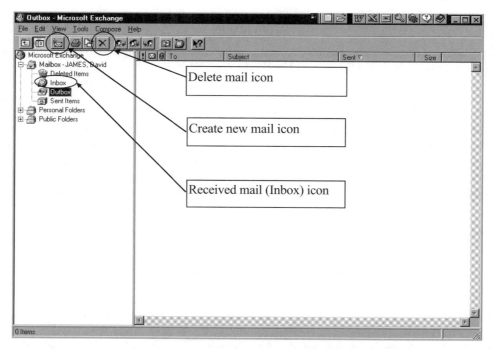

Figure 9.12   An example of an electronic mail application. Microsoft Exchange® © 1995 Microsoft Corporation. Used with permission (Microsoft Exchange is a trademark of Microsoft)

or get involved in debates. Email discussion groups are likely to become a large part of the independent component of taught modules in the future. Additionally, email skills will often be important to employers and will always be transferable.

ACTION POINTS

1   What is the difference between menu-driven and command-driven software?
2   Where on a computer might documents be stored?
3   What is the generic name given to software?
4   What type of application would be used to produce a scientific report?
5   Give three uses for a spreadsheet application in the study of exercise and sport.
6   Give an example of how a database might be used.
7   What types of visual aids can be produced using a presentation application?
8   Where are the data entered within a data analysis application?
9   Which three types of application allow the production of graphs?
10 Why might a graphics application be useful in the study of exercise and sport?
11 Explain why information and communication technology are inseparable.
12 Which two communication facilities does the Internet support?
13 Give the format of a web page address.
14 What is a search tool?
15 Why might an email discussion group be useful for students of exercise and sport?

Conclusion

Technology is now widely available to enable communication of information in addition to storage and retrieval of information. The World Wide Web (WWW) is the largest source of digital information, and is rapidly becoming more important for everyday life. The study of exercise and sport is made easier through access to the Internet, in terms of both information on the WWW, and the use of electronic mail (email). All functions performed on a personal computer are now menu driven, so it is rarely necessary to learn specific commands. Software applications are similar in their appearance, regardless of their specific purpose. A range of applications are available for use in the study of exercise and sport, including ones for word processing, spreadsheets, databases, presentation, data analysis and graphics. Skills developed using such applications will always be transferable, and are much desired by employers.

KEY POINTS

- Most personal computers now include a modem that enables communication with other computers via telephone lines.
- The Internet provides access to both the World Wide Web and electronic mail.
- Students of exercise and sport are expected to be familiar with a range of software applications.
- Skill development through the use of applications is transferable to the workplace.
- Equipment in an exercise and sport laboratory is often linked to a personal computer.
- Data analysis in exercise and sport always utilises computer applications.
- Report writing is facilitated and enhanced by the use of a word processing application.

Useful sport and exercise Internet addresses

British Association of Sport and Exercise Sciences: http://www.bases.co.uk/
European College of Sport Science: http://www.dshs-koeln.de/ecss/
International Sports Medicine Federation: http://www.fims.org/
Journal of Sport Sciences: http://www.journals.routledge.com/jss.html
National Coaching Foundation: http://www.ncf.org.uk/
National Sports Medicine Institute: http://www.nsmi.org.uk/

10

REPORT WRITING

AIMS OF THE CHAPTER

This chapter aims to provide an understanding of the importance of report writing in the study of exercise and sport. After reading this chapter you should be able to:

- explain the rationale for a scientific approach to report writing in sport and exercise;
- describe the various sections which normally make up a scientific report;
- provide detail on the content of each section of a report;
- apply scientific report writing to the study of exercise and sport.

Introduction

When studying exercise and sport, it is important to be able to plan, perform and write up experiments. This chapter will focus on writing scientific reports, and the sections that should be included within them. In addition to a description of how to write a report, a rationale is provided for the approach normally taken. Whether writing up a chemistry experiment or an exercise and sport experiment, the process will be similar. The approach to report writing is normally the same irrespective of whether you are an undergraduate student writing up a laboratory report or a professor writing a paper for publication. The skills developed through writing up scientific experiments are transferable to other report writing tasks.

One of the most important written pieces of work to be completed by an undergraduate student in an exercise and sport subject is the **dissertation**. For this piece of work report writing skills will be critical to success, so these skills should be correctly developed from an early stage in the course. An opportunity to develop these skills will be provided through the writing up of laboratory-based practical sessions. The development of report writing skills is a key part of academic work. The most common medium through which academic work is presented is concise reports, which are published in refereed academic journals. Thomas and Nelson (1996) provide a chapter on writing a research report, including detail on results and discussion sections, and ways of reporting research (pp. 385–424).

Scientific investigation of any subject is based on certain fundamental principles that have been covered in both Chapters 1 and 8. Research normally aims to test whether a theory stands up to rigorous investigation. The main research question is often posed in the form of clear hypotheses that can be tested. The findings of the investigation will dictate whether or not the hypotheses are accepted or rejected. Often statistical techniques are used to analyse the results in order to allow acceptance or rejection of the hypotheses with a degree of confidence. It will be explained how the fundamental principles of scientific investigations are accounted for in each section of a report. For information about the qualitative research process, see Graziano and Rawlin (1993) and Silverman (1993).

Writing style

One area of scientific report writing which many students find difficult is the normal convention of writing in the **third person**. Although it may seem natural to write in the first person, especially in the methods section of the report, convention dictates that the third person should be used. The use of the third person should help an author to report details in an objective form. In the following examples, the first paragraph illustrates the use of the first person and the second shows the use of the preferred third person.

Writing in the first person:

> In *our* last lecture, Prof. Harris told *us* that vertical jump height was an indirect way of assessing very short term power output. He also told *me* at the end of the lecture that *I* would have to be the participant in the following laboratory practical. The practical demonstrated to *us* that power determined from vertical jump height and power directly determined from a force plate differed slightly in the same participant.

Writing in the third person:

> In a previous lecture, Prof. Harris explained that vertical jump height was an indirect way of assessing very short term power output. At the end of the lecture, it was also explained that this author would have to be the participant in the following laboratory practical. The practical demonstrated that power determined from vertical jump height and power directly determined from a force plate differed slightly in the same participant.

From the above examples it is possible to observe the differences between the two styles, and see how the text written in the third person appears more objective, and less ambiguous. Reports are also written in the past tense (as is the case in the above example). Further discussion of this point is included in the 'Writing the methods' section, since this is an area of report writing which many students have difficulty with.

The importance of concise, clear writing in scientific reports cannot be over-emphasised. For the reader to extract information efficiently the inclusion and arrangement of sections, and sub-sections, within the report must be logical. As you will see in the 'Writing the methods' section below, this can often be a complicated task that requires practice. In many of the careers that a student of exercise and sport may move into, report writing will be an important transferable skill. A report should always be clear and concise, regardless of who the report is intended for.

Sections

A scientific report is normally written with clearly defined sections to lead the reader through all the necessary information. Without such sections, the reader would quickly become confused. Typical sections within a report include the following:

- Abstract
- Introduction
- Review of literature
- Methods
- Results
- Discussion
- Conclusion
- References.

A reader who wants to extract quickly certain information from a report may also find sections useful. For example, an **abstract** will normally provide the reader with a summary of the method and main results, and a brief concluding sentence. Sometimes, a quick look at the results section will provide the reader with the main findings of a study. The reader will then have to interpret the results themselves. Often a reader who is very familiar with an area of work will be in a good position to interpret the results, without referring to the author's interpretation.

It is advisable to read some published scientific reports to familiarise yourself with the content of these sections. A visit to a library that stocks journals relating to exercise and sport would be useful, to allow you to see some example reports in this area. Examples of refereed academic journals in exercise and sport are shown in Appendix 4. In the main, full papers are not presently published on the Internet; however, the first fully on-line journal in exercise physiology has recently been launched, and can be found at:

http://www.css.edu/users/tboone2/asep/fldr/fldr.htm

Visiting such a site will enable familiarisation with the format of a report in the area of exercise and sport.

When examining the journal contents, be careful to distinguish between original research papers and review papers. For the purpose of this chapter, original research papers should primarily be consulted. However, review papers can be helpful for showing how to write a good **review of literature** section (see, for example, the review papers by Hopkins 1991; Lehmann *et al.* 1993; Williams 1997). Some journals have two types of original research paper: normal papers and rapid communications. Rapid communications are shorter than normal papers, allowing the authors to get important findings into print quickly. A rapid communication is usually written in a form similar to that of a laboratory report from an undergraduate student. It is therefore worth examining such reports to observe how the authors manage to include so much information in such a concise manner.

Writing the abstract

Often, when there is no time to read a complete paper, the reader will look just at the abstract, to obtain a summary of the information included in the article. This is why it is important to learn how to incorporate all the necessary information into the abstract. An abstract format is not only included as a section of an original research paper, but is also widely used to inform delegates at a conference about the content of a presentation. Many journals in sport and exercise make use of abstracts to provide Internet users with information on papers. It is possible to examine a range of abstracts on the Internet. The following address may be a good place to start:

http://link.springer.de/link/service/journals/00421/tocs.htm

It may also be useful to look at an edition of a journal which provides abstracts from conferences. By studying such a publication, slightly longer abstracts may be viewed, which demonstrate the approach taken for a conference. An example is the *Journal of Sport Sciences*, volume 16, number 5, which includes abstracts from the 1997 annual congress of the European College of Sport Sciences.

Some journals specify the content of the abstract, whilst others allow the reader more freedom. However, in general, an abstract will include an introductory and a concluding sentence, along with a description of the method and the main results. The following detail is normally required:

- Single sentence to introduce the problem
- Aim of investigation, including hypotheses
- Method:
  - Subjects
  - Protocol
  - Measurements
  - Equipment
  - Analysis of data
- Results summary
- Conclusion in relation to introductory sentence and hypotheses.

A limit of between 150 and 250 words is often applied to an abstract. Normally dissertation abstracts are slightly longer, whilst journal paper abstracts are shorter (closer to 150 words).

Writing the introduction

The purpose of an **introduction** is to allow the reader to understand why the investigation has taken place. The aim of the introduction is to:

1 Enable the reader to place the investigation in context.
2 Provide a rationale for the investigation.
3 Delimit the investigation.
4 State the aim of the investigation.
5 State the hypotheses.

The content of the introduction may differ depending on the application. Often, an introduction in a journal paper will also review the literature. However, in most cases, a separate introduction and review of literature will be included. If there is a separate literature review, the introduction is the section that takes the reader from a very general focus to the specific aims of the study. For example, an undergraduate biomechanics laboratory report included the following in its introduction:

Performance in many sports depends on the ability to generate power over a very short time period. Often short term power production is used to raise the performer in a vertical plane against the force due to gravity… In order to assess the difference in power production capability of performers in different sports, vertical jump performance was determined in six class members. Three of the participants were basketball players, and three were cyclists.

In the above extract, the author explained the general context of the study, before explaining to the reader exactly what would be investigated. An introduction helps the reader to contextualise the investigation. Once the study has been placed in context, the **hypotheses** may be stated. In most scientific reports, the hypotheses are stated in the form of open questions. For example:

The aim of this investigation is to determine whether a prior fatiguing bout of running exercise alters gait during sub-maximal running.

In a dissertation, however, you would normally be required to formally state a null hypothesis and an alternative hypothesis. For example:

- Null hypothesis (H_0): A prior fatiguing bout of running exercise does not alter gait during sub-maximal running.
- Alternative hypothesis (H_1): A prior fatiguing bout of running exercise alters gait during sub-maximal running.

Sometimes it is possible to specify the direction of the change in the measured variables within the hypothesis. For example:

- H_0: A prior fatiguing bout of running exercise does not increase the duration of the stance phase during sub-maximal running.
- H_1: A prior fatiguing bout of running exercise does increase the duration of the stance phase during sub-maximal running.

For further detail on the formulation of hypotheses, and the association with the choice of statistical test, see Chapter 8.

It should be noted that the introduction does not always include the research question and hypotheses. Sometimes the research question and hypotheses are included after the review of literature.

It is common to have statements in the introduction supported by **references**, even if the introduction is to be followed by a review of literature. Factual sentences should be accompanied by reference citations which support the statements (see 'Producing the references' section for details). If, however, the factual information is the author's own thought, then it should be acknowledged that it is a personal comment. For example:

In the experience of this author, it has been observed that many high level athletes take the area of nutrition very seriously.

Writing a review of literature

The distinction between the introduction and review of literature is one that many students find unclear. Whereas an introduction generally contains references, they are normally for general points of fact. It is in the review of literature that the author makes reference to any studies that have been performed previously in the area, and the results of these studies. So, for example, in the introduction the following statement might appear, supported by a reference(s):

> Endurance performance is predicted by a range of physiological and biochemical parameters, both at the whole body and cellular level (Jones, 1997).

By contrast, in the review of literature, a reference may be used in the following way:

> In a study that examined eight well-trained cyclists, maximal minute power produced during an incremental cycling test to exhaustion predicted 60% of the inter-subject variation in cycling time trial performance (Smith *et al.*, 1996).

The skill of writing a good review of literature is in linking the discussion of previous studies together in a logical order. Even if your study is to be the first in a particular area, it is appropriate to make reference to previous work in related areas. Fortunately, in the study of exercise and sport most of the techniques available for student work have been widely used previously, thus there is usually no shortage of literature to which to make reference.

Previous research should be referred to objectively. As the author of a report, it is important to write the review of literature as though the study had not been carried out, even if it has. By taking this approach, the findings of the study should not influence the review of literature. It is important to take this prospective approach, since reviewing the literature, in addition to the investigator's interests, should influence a study. Only once the literature in the area of interest has been reviewed will the investigator be in an informed position to design the study. In practice, many researchers will read the literature, then plan and execute the study, and then write the review of literature along with the other sections of the report. By taking a prospective approach to writing the review, the literature will be reported as it was interpreted prior to the study. It is within the discussion that the findings of the study are related to the findings reported within the review of literature.

For examples of literature reviews have a look at review papers in refereed scientific journals. The journal *Sports Medicine* consists mainly of reviews; the contents and abstracts of this journal can be found at:

> http://www.adis.com/journals/sportsmedicine/

This type of journal is a good place to start a literature search for a project, as well as for finding out about writing a literature review. For detailed discussion of referencing techniques see the 'Producing the references' section below.

Writing the methods

The description of how a study was carried out is usually referred to as the **methods** section.

A frequently asked question is, why is it important to describe the study – surely it is only the results of a study that are important? However, there are two important reasons why the report of a study should contain a methods section:

1   Scientists may wish to repeat exactly the study to confirm the findings.
2   The method employed may influence the results dramatically.

It is quite often necessary to repeat a study, either in an identical way, or by changing part of the method slightly. For example, it may be important to replicate the study exactly, as mentioned earlier, in order to add information gained from additional subjects. It may also be important to replicate the study exactly, but to use different subjects. The findings of the experiment may then become generalisable to a wider population.

There are numerous ways in which a slight change in the methods may influence the findings of a study. This point is best illustrated with an example: In one study in which the effects of a carbohydrate drink on exercise performance were to be assessed, participants in the experimental group consumed a carbohydrate drink and participants in the control group consumed no drink (study 1). In another study the experimental group participants consumed a carbohydrate drink, and the control group participants consumed an equivalent amount of water (study 2). When the studies were examined, it was found that the experimental group performed significantly better than the control group in study 1 but not in study 2. The only difference in the study design was that water was consumed in the control condition in study 2, but not in study 1. This difference in methods is important, since it has demonstrated that, under the experimental conditions, a carbohydrate drink is better than no drink, but not significantly better than water in the same quantity. The conclusion from this may be that the water in the carbohydrate drink is eliciting a large part of the performance improvement, rather than the carbohydrate *per se*. This example demonstrates the importance of a detailed methods section.

When the literature in the area of sport and exercise is examined, what appear to be similar studies often have slightly different methods. A well-read student of sport and exercise will be able to differentiate between methods, and make judgements about the likely effect upon the results of the study.

In order to produce a clear and coherent methods section it is essential that the section is well planned, and structured appropriately. It is important to consider the following:

- the usual conventions for describing a study;
- the necessary considerations when using human subjects;
- the correct terminology and style of writing.

In scientific writing, the methods section of a study is usually written in a standardised way. This general format is the same whether it is a sport and exercise study or a physics study that is being reported. The methods usually contain detail on the following:

- participants used in the experiment;
- the general protocol;
- the specific measurement details;
- the analysis of the data;

In the subject of sport and exercise, most of the studies performed will involve the use of **human participants**. Details on the participants are usually given in the methods section, and may include descriptions of the population they have been selected from, the process of selection, and their physical attributes. Very strict guidelines exist about the treatment of human participants when they are involved in a scientific study (see Chapter 1), and the methods section of the study report should make reference to the fact that such guidelines have been adhered to. Although an undergraduate student may not need to be concerned with abiding by the guidelines of academic journals relating to the treatment of human participants, even in undergraduate reports of laboratory practical work it is good practice to make reference to participant treatment. Normally, it is reassuring to know that undergraduate laboratory practical sessions only include procedures that have been previously cleared with the institution's **ethics committee**. For further discussion of ethical issues and **informed consent**, see Chapter 1.

An example of how the participant section of your methods might read is:

Eight well-trained female cyclists (mean ± SD: body mass, 64.1 ± 1.5 kg; height, 1.68 ± 0.11 m; $\dot{V}O_2$max, 4.11 ± 0.27 L·min^{-1}) agreed to take part in the project which was approved by the Institutional ethics committee. All subjects gave written informed consent to take part in the study following completion of a health screening questionnaire.

It is usual to give details about groups of participants at the end of the participant description if the experimental protocol requires the participants to be split into more than one group. The reader can then immediately see whether the groups are matched or similar.

The importance of an accurate description of the protocol used to examine a particular issue has already been discussed. The question remains, how can the protocol be reported in an accurate, objective and unambiguous way? Accuracy will be achieved if every important part of the study is reported in detail. It is also necessary that the general layout of the study description assists with the understanding and piecing together of information. The key to a good methods section lies in the organisation of the information for the benefit of the reader.

A useful way to think about this process is to consider the most general aspects first, then gradually move into the fine detail. Also, describe the protocol in chronological order (i.e. in order of time). For example, you might start with something like:

Participants visited the laboratory on three occasions spaced one week apart. On the first occasion, subjects had their mass, height and fat percentage determined. On the second occasion, ...,

and may finish with the following information:

Immediately after the second visit to the laboratory, participants were instructed to consume one litre of a 6% carbohydrate solution....

Notice in the preceding example that the information is presented objectively. This means that the protocol of the study is presented factually and as it happened, with no attempt by the researcher to interpret what happened. The presentation of information in this way helps to avoid a situation where the information presented might be ambiguous or misleading. Many students find this aspect of the methods section particularly difficult. The following provides an example of how a less objective description might lead to ambiguity:

> The subjects were asked to visit the laboratory on three occasions, and these visits were separated by a few days. When the subjects came to the laboratory, I did several tests, some of which were repeated on more than one occasion. On the first occasion the subjects were asked to stand on the scales to measure their weight.

In this example the more subjective style of writing makes the description less clear, and incorporates the experimenter's interpretation of what happened, rather than giving an exact record of events. Also, the use of the first person in places makes the description less easy to follow.

Once the detail of the protocols has been provided, it is then appropriate to describe the way in which measurement took place. It is usual in a study for several different measurements to be taken, and for some of the measurements to be repeated during different parts of the protocol. For example, heart rate may be measured in all exercise tests in a study, but the method of measurement need be described only once. The description should usually include precise detail, so that exactly the same procedure of measurement may be repeated on a subsequent occasion. For example, if a heart rate monitor is used to measure heart rate throughout a study, the description might be as follows:

> Heart rate was determined via telemetry (Model HR121, Manufacturer's name, Manufacturer's location). Prior to exercise, subjects placed an elasticated strap around their chest, which was tightened but still allowed unrestricted chest movement. The subjects also placed a watch on their wrist, which stored the information transmitted from the strap at five second intervals. On termination of exercise, the stored information was downloaded to a computer for subsequent analysis.

In the above example, enough information is given about the equipment to enable another scientist interested in measuring the same phenomenon to purchase it. The method of measurement, and equipment used to record the measurement, is usually followed by the necessary information in brackets. In the above example, the model number of, and name and location of the manufacturer of, the heart rate monitor used are provided. A simple example has been given here, but some of the equipment you may use may be far more complicated. An example of this type is provided in Application 2.

The part of the methods section that describes the data analysis should clearly explain how the investigator has treated the raw data. The raw data is simply the information gathered as measurements are undertaken. This information is invariably difficult to interpret in its raw form. Some analysis has to take place before a reader can interpret the information. At a fundamental level of data analysis, some calculations may have to take place to move from basic variables to derived variables. By this we mean that two or more measurements may have to be used to calculate a further variable. A simple example of this would be the use of information about the distance covered by a runner and the time taken to cover the distance. With these two pieces of information, speed can be derived (i.e. distance divided by time). The clear description of such a process is very important, since the analysis of data in different ways can lead to different findings.

A further area to be covered in the data analysis section of the methods is the description of the statistical techniques that have been used. Again, the choice of statistical procedure can markedly affect the study findings. The appropriate methods for analysing various types of data are discussed in Chapter 8.

A large amount of variation is present in the format of, and detail provided in, the methods section. Part of the reason for such differences lies with the requirements of the particular institution where the study has been conducted, or of the outlet for publication of a study. If you are an undergraduate student, the supervisor of your laboratory practical sessions may provide specific guidelines for the layout of the methods section of written reports. It is, however, a good idea to have a look at some examples of methods sections in the published literature. Chapter 4 in Thomas and Nelson (1996) provides comprehensive further reading on formulating the method. Part II of Hyllegard *et al.* (1996) provides a more detailed insight into the formulation of appropriate methods, with pp. 116–118 providing an entertaining analogy.

Writing the results

For appropriate presentation of **results**, this section should be read in conjunction with the chapter on data analysis (Chapter 8). Whilst Chapter 8 provided the information necessary to deal with various types of raw data, this section will enable presentation of the main findings in the best format.

Normally, a results section of a scientific report will include the following:

- a description of the participant characteristics;
- tables and figures of results;
- a description of the main findings.

Notice that no interpretation of the results is attempted in this section. Results should be interpreted only in the discussion section.

It is important to report the results in an objective form, and in the **third person**. When the results are reported in this way, the reader is able to obtain the main findings of the study quickly, and interpret the findings themselves if necessary.

In order for information to be extracted quickly by the reader, the results should be presented in order of events, if possible. As was the case with the methods section, ordering the presented information chronologically will aid the reader.

The description of the participant characteristics, which is usually the most appropriate starting point, will often be presented in the form of a table. The text which may accompany such a table may make reference to the type of participants in the group. Information about the degree of similarity (homogeneity) or difference (heterogeneity) of the group, and the training status of the group, may be included. It is often necessary at this point to mention if any participant had to withdraw from the study. If the study required a division of participants into two or more groups, the details of each group would normally be presented. It is important not to repeat information in the text that has been supplied in a table.

The main findings of the study will often be presented in a table or in graphical form. The old saying 'A picture [i.e. table or figure] can tell a thousand words' is relevant here, but only if the information is presented in an appropriate form. Before producing tables and figures it is worth spending some time examining good practice (for further information see Thomas and Nelson 1996: 399).

The term 'figure' may refer to a graph, a schematic, or a picture. A schematic is often used to present information that is difficult to explain in text, due to the inter-relationships between pieces of information. Figure 8.1 is a good example. A picture is used less frequently in scientific reports, but one may be included to illustrate a certain piece of novel equipment. For the purpose of this section on writing the results, only graphs will be considered (along with tables), since schematics and pictures are usually included in other sections of a scientific report.

The first question is 'What information should be presented as a table and what information should be presented as a graph?'. Generally, graphs are the presentation form used when data are plotted in relation to a continuous variable, for example 'time' with a unit of seconds or minutes. Also, graphs are usually used when comparing two or more groups of data. Tables tend to be used when data relate to a discontinuous variable, for example subject number. In order to get a feel for the sort of information that is normally included in either a graph or a table, have a look at some scientific reports in exercise and sport journals. In addition, Chapter 8 of this book includes several examples of different types of tables and graphs.

Tables and figures must be properly titled. The title (sometimes referred to as the legend) should be such that the table or figure can be understood without reference to the text. Clearly, therefore, the title has to provide a substantial amount of information in a concise form. In addition, convention dictates that the title for a figure is placed at the bottom of the figure, whereas the title for a table is placed at the top of the table (see Table 10.1 and Figures 10.1–10.5 for examples).

The column headings of a table and axes of a graph must be properly labelled. The label should describe what is on the axis or in the column, and provide the unit of measurement.

Table 10.1 Individual and group training, anthropometric and physiological characteristics for eight well-trained runners [An example of a table]

Participant number	Age (years)	Training history (years)	Mass (kg)	$\dot{V}O_2max$ (L·min⁻¹)
1	40	7	59	4.11
2	31	10	68	5.08
3	35	12	79	4.87
4	47	3	72	3.99
5	29	6	67	3.76
6	36	8	65	4.21
7	35	13	76	4.65
8	32	11	74	4.86
$\bar{x}$	35	9	72	4.49
SD	6	4	5	0.50

Mass, body mass; $\dot{V}O_2max$, maximal oxygen uptake determined during incremental treadmill running.
$\bar{x}$, group mean average; SD, group standard deviation.

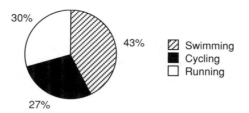

Figure 10.1 Proportion of participants who consider either the swimming, cycling or running event the most important in an Olympic distance triathlon [An example of a pie chart]

Failure to include an appropriate unit is a common reason for marks being deducted from a student's work.

It is preferable to avoid the inclusion of vertical lines in a table. It is better if the text/numbers in each column are simply divided by a tab (see Table 10.1). It is important that the information in each column is aligned. Sometimes the information is aligned to the left of the column, and sometimes to the right. The advantage of aligning numbers to the right of a column is that, providing the number of decimal places is consistent within the column, any decimal points will then line up.

Horizontal lines are included in tables to isolate the column heading from the content of a column, and to indicate the end of the table. The title of a table usually sits above the first horizontal line and any additional notes or explanations are usually placed below the final horizontal line. If a table contains information about individual subjects and also group data, a horizontal line is normally included to differentiate the information (see Table 10.1).

Since numerous types of graph are found in scientific reports, only a brief explanation and a few examples will be provided for each type. Further examples can be found in published articles in refereed exercise and sport journals.

A **pie chart** is a simple way of presenting data that are categorised (see Figure 10.1). A pie chart simply expresses the proportion of data that falls within each category as a section of a pie. It is rare to see a pie chart in the sport and exercise literature, since such information is more typically expressed in a percentage form in a table. The reason that such information is tabulated rather than expressed in a graphical form is that a graphical representation does not always help the reader to understand the information any better than a table.

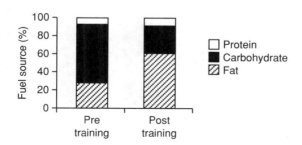

Figure 10.2 Proportion of protein, carbohydrate and fat sources used during exercise prior to and following a training intervention [An example of a column chart]

Sometimes the same participants may be monitored prior to and following an intervention, or more than one group may be considered. If the data are still categorised, a column chart may be used (Figure 10.2).

A variety of **bar charts** (sometimes referred to as **histograms**) are found in the scientific literature (Figure 10.3). Bar charts do not differ substantially from the column charts described above. Bar charts are often used if the same measurement is made on several occasions in a group of participants.

If the same measurement is taken from a number of participants on a number of occasions, and the participants are from two or more groups, it is often better to view the results as a **line graph** (Figure 10.4). However, care must be taken when joining data points by lines. It is appropriate to join points by lines only if points are related in some way, for example the same measurement has been made at different times. Care must also be taken when selecting which variable the x-axis (abscissa) and which forms the y-axis (ordinate). The general rule is that the independent variable forms the x-axis and the dependent variable(s) forms the y-axis. The independent variable refers to the variable manipulated by the experimenter in order to observe the change in the dependent variable.

If the points on a graph simply refer to individual participants, and demonstrate the relationship between two variables, a **scatter plot** may be produced. The same rules for allocating variables to axes as for the line graph apply to scatter plots.

The text that accompanies the tables and figures should draw the reader's attention to important information, and make reference to the overall findings. For example, if all participants except one responded in a certain way to an intervention, attention may be drawn to the unique result. The order of presentation of the main findings should be carefully considered. It is often appropriate to take the reader through the results as they emerged from the study.

The presentation of statistical results, whether descriptive or inferential in nature, should include an explanation of abbreviations used. For example, the first time that a sample mean average is mentioned, it is important to write the term in full with the abbreviation in brackets following, i.e. 'mean ($\bar{x}$)'. When an inferential test has been performed to establish whether a difference exists between two data sets, the confidence level selected for the test is normally

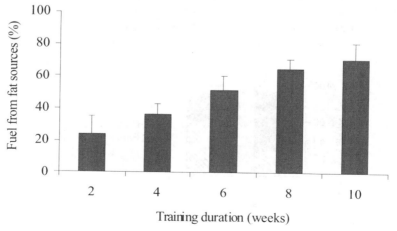

Figure 10.3 Proportion of fat used during exercise after 2, 4, 6, 8 and 10 weeks of training (mean and standard deviation) [An example of a bar chart, or histogram]

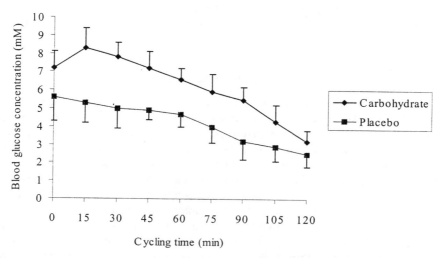

Figure 10.4 Blood glucose concentration during 120 min of heavy exercise following ingestion of either 100 g of carbohydrate (10% solution) or placebo 1 hour prior to exercise (mean and standard deviation) [An example of a line graph]

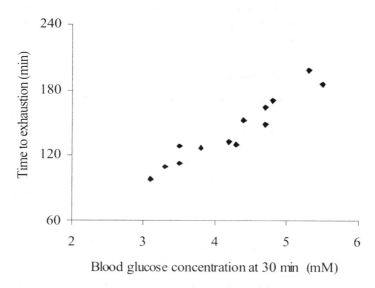

Figure 10.5 Relationship between time to exhaustion and blood glucose concentration after 30 min of exercise [An example of a scatter plot]

included after a statement (e.g. 'a significant difference existed between groups 1 and 2 ($P <$ 0.01)'). In some situations it may be appropriate to provide the test statistic. For example, in a t-test the test statistic would be given as $t = 4.54$, where 4.54 would refer to the t-value derived from the statistical test. For further information regarding data analysis see Chapter 8.

Writing the discussion

Once the results have been presented, it is necessary to explain the results to the reader. The explanation should make reference to the information provided in the introduction, particularly the hypotheses, and in the review of literature. Any interpretation of the results should refer to the findings of previous investigations in the area. It is often necessary to include further references, beyond those mentioned in the review of literature, to support points that you, as the author, are attempting to make. Writing a good **discussion** is a difficult skill to learn, and looking at discussion sections of published articles in refereed journals may be helpful.

As with the other sections of the report, it is important to lead the reader through the discussion in a logical order. It is often useful to start by discussing the participant group recruited for the study, and how the group relates to the participants in previous studies. The implications of the trained status of the group, or whether the participants are female or male or mixed, should be discussed.

If inferential data analysis has been undertaken, it may be helpful to start talking about the variables that changed significantly within the study. The changes in these variables, in relation to changes in other variables that did not reach a significant level, may also be of interest. At some point the **hypotheses** stated in the introduction (or after the review of literature) should be accepted or rejected in light of the findings of the study.

Writing the conclusion

The main objective of the **conclusion** is to relate the main findings to the aims of the study and the hypotheses. Were the aims of the study realised, and if so, what were the main implications of these findings? It is quite common to end the conclusion section with a mention of future research which leads on from the present study. It is always useful to state clearly and reasonably forcefully the implications of the findings, since this highlights the justification for the study. Having said this, it is worth exercising some caution at this stage: it is a common mistake to make claims about the implications that are too ambitious, or are not substantiated by the findings of the study.

Producing the references

A common question from students of sport and exercise is, why is referencing so important? One of the most important reasons is that if the work of other investigators is not acknowledged, a serious academic offence has been committed. This offence is known as **plagiarism**. In many academic courses the penalty for plagiarism is a complete loss of marks for the piece of work in question, or more seriously a loss of marks for the whole module or semester. It is not unknown for students to fail their degree for plagiarism. A related offence with similar consequences is **collusion**, which is when the work of a collaborator is not acknowledged. For example, if two students worked together on a project but one student attempted to gain full credit for the work, this would be collusion. If an academic collaborates with another individual, the input of that other individual must be fully acknowledged.

From the outset, it is important to develop a logical system for listing sources that have been consulted. In many cases the sources are numerous, so a tidy method of listing them is essential. If a student learns a recognised style of referencing from early in their studies, their work will benefit in the long term. Although the methods used in the most common referencing systems may not seem logical or simple initially, their long standing use by academics is evidence of their usefulness.

There are two components to a referencing system:

1   Inclusion of reference citations in the main text of the report.
2   Production of a list of references at the end of the report.

The styles which are adopted by most institutions and refereed academic journals are the Harvard and **Vancouver** referencing systems. With the Harvard system the references appear in the reference list in alphabetical order. With the Vancouver system the references appear in the reference list in the order in which they were cited within the text. For the Harvard system, the citations within the text are given in the form of author name(s) followed by the year of publication, e.g. Williams and James (1999). When there are more than two authors of a publication, the reference is given as Williams *et al.* (1999). For the Vancouver system, within the text a number is given to identify the reference in the reference list at the end of the report.

Both the Harvard and Vancouver systems are widely used in refereed academic journals in the area of sport and exercise, and these journals should be consulted for examples. Both systems dictate a preferred convention for style and format of the references in the list at the end of the report. Examples for a journal article, a book and a chapter in an edited book are given below.

- Journal article:
Atkinson, G. and Nevill, A.M. (1998) Statistical methods for assessing measurement error (reliability) in variables relevant to sports medicine. *Sports Medicine* 26(4): 217–238.
- Book:
Hinton, P.R. (1996) *Statistics Explained: A Guide for Social Science Students*. London: Routledge.
- Chapter in an edited book:
Rowbottom, D.G., Keast, D. and Morton, A.R. (1998) Monitoring and preventing overreaching and overtraining in endurance athletes. In *Overtraining in Sport*, R.B. Kreider, A.C. Fry and M.L. O'Toole (eds). Leeds, UK: Human Kinetics.

APPLICATION OF SCIENCE TO EXERCISE AND SPORT

Application 1

During many exercise and sport courses students will be expected to perform a laboratory practical following a lecture. The aim of the practical will normally be to provide the student with a greater understanding of points covered in the lecture. A laboratory practical is usually planned by the lecturer, and the students are provided with a protocol to follow. Therefore, the main learning is taking place through the observation of, and involvement in, the practical, and the comprehension and understanding of the experiment through the production of a report.

A student is forced to think through many issues covered in the practical when trying to express the main points in a clear, objective manner in a scientific report. Production of an introduction and review of literature section will require the student to look at some of the recommended reading. Through such reading, the student will learn about the basis for the practical, and the outcome of other attempts to perform the same practical. When describing the methods used, the student will learn to explain the participants involved, protocol employed, and the treatment of collected data clearly and objectively. Through production of the results section, the student will learn appropriate techniques for analysing and presenting the results. The discussion and conclusion will provide experience of relating the findings to previous studies, and explaining the practical implications.

Application 2

The production of the methods section will become more complicated if apparatus that is formed from several pieces of equipment has been used in the study. An example would be apparatus used for examining gas exchange (see Chapter 2 for further information). This apparatus is often used in studies of exercise and sport; as it consists of many pieces of equipment, the description must be carefully written. It may be useful to structure the description of equipment in a chronological order, as is done for other areas of the methods section. For example, when examining pulmonary gas exchange the equipment used could be described in order of where the gas moves after exhalation. You might start with the mouth-piece, then mention any connecting hoses, followed by collection or measurement equipment. Similarly, for the measurement equipment used for the expired gas, the numerous pieces of equipment should be treated in a logical order. Finally, it may also be important to describe calibration procedures for each piece of measurement equipment. The principle of calibration is covered in Chapter 2.

ACTION POINTS

1 What are the various sections normally included in a scientific report?
2 List five details that should be included in the abstract section of a scientific report.
3 What are the four areas to cover in the methods section of a scientific report?
4 What are the three areas to cover in the results section of a scientific report?
5 Describe the main objective of the conclusion section of a scientific report.
6 What referencing styles are normally used in the study of exercise and sport?
7 Explain how a book and a journal paper are referenced in the list at the end of a report (bibliography).
8 In the Harvard system, how is a reference with two authors cited in the text of a report? In the same system, how is a reference with more than two authors cited?
9 In which section of a scientific report does the author provide the context of the study?
10 In which sections of a scientific report does the author make reference to previous research in the area?

Conclusion

It is essential that scientific report writing skills are developed during the study of exercise and sport, since a scientific report is the most common form of communication of information in this area. Most undergraduate courses in sport and exercise-related subjects have a final year dissertation based on a scientific investigation performed by the student. The dissertation is the most important single piece of work completed by the student during such courses. The development of scientific report writing skills throughout a course will enable the student

Table 10.2 A summary of the content of each section of a scientific report

Report section	*Content*
Abstract	Sentence to introduce the problem Aim of investigation, including hypotheses Brief description of the methods Summary of results Conclusion in relation to hypotheses
Introduction*	Narrows from general context to specific investigation rationale Contextual references provided Research question and/or hypotheses†
Review of literature	References provided for previous studies in the area Detail given on previous studies, including critical appraisal Research question and/or hypotheses†
Methods	Description of the study in the third person in an objective form, including: Participant details General protocol narrowing to greater detail Specific measurement details, including calibration procedures and equipment details Data analysis procedures
Results	Description of participant characteristics Tables and figures of results with informative titles Objective description of results with no attempt to interpret findings
Discussion	Explanation of results Comparison of results with findings from other studies Implications of results for practice
Conclusion	Relates findings to research question/hypotheses Requirements for further research Summary of implications for practice
References	List of all references included in the text Usually conform to Harvard or Vancouver systems

*The introduction may sometimes include the review of literature.
†Research questions and/or hypotheses may be included within either the introduction or the review of literature.

to claim an extremely useful transferable skill, in addition to getting a better result for their dissertation project.

A scientific report does not differ greatly across the range of scientific disciplines. The writing style should be objective, and the writing should be in the third person. The presentation of information should enable the reader to extract the necessary information quickly (see Table 10.2). The use of primarily factual material, presented in a logical order, will help the reader to extract the important information. The factual information should, however, be presented in relation to other previously published work in the area.

The nature of scientific work is based upon the idea that a testable scientific theory is developed, which is then subjected to rigorous appraisal. If the theory is found to be inadequate then it is adapted. Therefore any scientific investigation should have clear hypotheses, based on theoretical underpinnings, which can be tested. Once the results of the study are known, the hypotheses are either accepted or rejected. Since this approach underpins all scientific work, it is essential that any scientific reports present clear, testable hypotheses. The conclusion of a scientific report should always make reference to the hypotheses.

KEY POINTS

- Scientific reports are usually divided into sections, to allow the reader to extract information quickly.
- The use of sections in scientific reports is also helpful to the author when writing the report.
- Usual sections included in a scientific report are an abstract, introduction, review of literature, methods, results, discussion, conclusion, and references.
- An abstract normally includes an introductory sentence, a description of the method, and main findings, with a concluding sentence.
- An introduction normally places the investigation in context, and provides detail on the aims and testable hypotheses.
- A review of literature normally describes previous studies in the area in an order which helps the reader to follow a logical thought process towards the research question.
- A methods section normally includes a description of the investigation, including consideration of the participant, protocol, measurement, equipment, and data analysis details.
- Appropriate style of writing and use of terminology will make the description of the methods objective and unambiguous. As a consequence, it should be possible to repeat the experiment exactly if required.
- A results section will normally present the results in the most appropriate manner, with no form of interpretation.
- A discussion section will normally consider the results in relation to the hypotheses, and to the results of previous studies in the area.
- A conclusion section will normally make reference to the aims of the study, and state whether the hypothesis was accepted or rejected.
- Referencing should normally conform to the Harvard or Vancouver systems.

Bibliography

Graziano, A.M. and Rawlin, M.L. (1993) *Research Methods: A Process of Enquiry*. New York: HarperCollins College Publishers.

Hopkins, W.G. (1991) Quantification of training in competitive sports. *Sports Medicine* 12(3): 161–183.

Hyllegard, R., Mood, D.P. and Morrow, J.R. (1996) *Interpreting Research in Sport and Exercise Science*. St Louis, MO: Mosby-Year Book.

Lehmann, M., Foster, C. and Keul, J. (1993) Overtraining in endurance athletes: a brief review. *Medicine and Science in Sport and Exercise* 25(7): 854–862.

Silverman, D. (1993) *Interpreting Qualitative Research*. London: Sage Publications Ltd.

Thomas, J.R. and Nelson, J.K. (1996) *Research Methods in Physical Activity*. Champaign, IL: Human Kinetics.

Williams, C. (1997) Children's and adolescents' anaerobic performance during cycle ergometry. *Sports Medicine* 24(4): 227–240.

Further reading

Beynon, R.J. (1993) *Postgraduate Study in Biological Sciences: A Researcher's Companion*. London: Portland Press.

Girden, E.R. (1996) *Evaluating Research Articles*. London: Sage Publications Ltd.

Phillips, E.M. (1993) *How to get a PhD: A Handbook for Students and their Supervisors*. Buckingham: Open University Press.

GLOSSARY

Abstract   An overview of a research paper or research report.

Acceleration   Increase in velocity with respect to time.

Acute   Short term or sudden.

Adenosine triphosphate   A high-energy compound essential for human energy transfer.

Application   In relation to computers, software with a particular function.

ATPS   This abbreviation refers to gas volumes under specific conditions, which include ambient temperature (in kelvins), ambient pressure (in mm of mercury) and saturated with water vapour (at 37°C water vapour pressure is 47.1 mmHg).

Bar chart   A graphical representation of data, in which bar height represents the magnitude of the dependent variable.

BMR   The basal metabolic rate (BMR) is a reflection of the body's heat production during basal conditions. It is indirectly determined by measuring whole body respiratory variables under specific conditions, which include no food consumption in the previous 12 hours, no intense physical activity in the previous 24 hours, and a 30 minute period of rest in the supine position prior to data collection.

BTPS   This abbreviation refers to gas volumes under specific conditions, which include body temperature (in kelvins, usually 310 K), ambient pressure (in mm of Mercury) and saturated with water vapour (at 37°C water vapour pressure is 47.1 mmHg). Pulmonary physiologists invariably report ventilation and lung function data in a BTPS form.

Calibration   A procedure comparing known reference values against measured values to determine accuracy.

Capacity   A magnitude of the ability to do something.

Carbohydrate   A nutrient supplying energy.

Case study   A study of a single instance or matter for discussion.

Category data   A data set of responses that allows a researcher to place the answers into a category, i.e. brand of running shoe is categorical data.

Chronic   Continuing for a long time or constantly recurring.

Coefficient of variation   A standardised measure of variation which is a ratio of the standard deviation and the mean average, usually expressed as a percentage.

Cohesion   The attractive forces which hold together atoms or molecules forming a solid or liquid.

Collusion   A secret act for a fraudulent purpose.

Compression   The application of forces that squeeze together.

Computer system   A system found on a computer that facilitates use of the computer functions, e.g. Microsoft Windows, Apple Macintosh systems.

Concentration   The strength of a solution, often expressed as the amount of solute for a given volume unit of solvent. The concentration can be expressed as weight in grams or milligrams, number of ions in moles (mol) or number of solute ions in equivalents (equiv.).

Conclusion   The summing up at the end of an essay or report.

Conduction   In relation to heat, the transfer of energy when two surfaces are in contact with one another. In relation to electrical signals, the transfer of electrical energy from one place to another.

Confidence interval   An interval calculated as a number of standard deviations from the mean average of a data set. Used to determine with a degree of confidence whether a sample comes from a population or an individual observation comes from a sample.

Control group   A group that does not receive any intervention or treatment in a study.

Correlation coefficient   A standardised measure of the degree of linearity of the relationship between two variables.

Cross-over design   A study design where the control group becomes the experimental group and the experimental group becomes the control group.

Cycle ergometer   A device for doing work on by rotating a wheel.

Data analysis application   Computer software which has a primary function of analysing data.

Database   An electronic or computer-based system for organising data or information.

Data   Information, sometimes in a numerical form, which may be raw or derived from raw data.

Declaration of Helsinki   A written document based on a meeting in Helsinki regarding the treatment of human subjects in clinical research.

Density   Density is a function of the ratio of mass and volume. The unit is $g \cdot cm^{-3}$.

Descriptive data analysis   Examination and manipulation of data in order to describe it.

Direct measurement   Examination of a phenomenon through direct means.

Discussion   In relation to a report or essay, a section which examines the results in relation to other work in the area.

Disk   In relation to computers, an information storage device in the form of a hard disk, floppy disk or Zip disk.

Dissertation   An extensive report of original research often forming part of a programme of study in academia.

Document   In relation to computers, the product of word processed information.

Drag   A force opposing movement.

Efficiency   A term relating energy use to work done.

Elasticity   The property of a material which allows it to return to its original shape after deformation.

Electronic mail   A message sent from one computer to another via an electronic network.

End-tidal   The end of expiration during a breathing cycle.

Energy   The capability to do work, expressed in units of joules (J) or kilocalories (kcal).

Energy expenditure   The energy expended by the body as a result of the resting metabolic rate and the thermic effects of food and physical activity.

Ethics committee   A group that has responsibility for examining the conduct of working practice to see that fair treatment of individuals takes place.

Experimental group   Participants in a study who receive the intervention.

Extension   In relation to joint angle, an increase in angle.

Fat   A nutrient composed of the chemical compounds carbon, hydrogen and oxygen, and which forms an essential part of the diet.

File In relation to computer systems, a structure for the storage of information.

Force plate A metal plate which is inserted into the ground so that its surface is level with the ground. The plate can determine force in one vertical and two horizontal directions through a series of strain gauges.

Force A vector quantity which causes acceleration or deceleration of an object when acting upon it. Force is measured in newtons (N).

Friction A resistive force when one object moves relative to another object and their surfaces are in contact.

Health questionnaire A series of written questions which ask individuals about their health and from which, for example, decisions can be made about safe exercise intensities.

Health screening A process of examining health status prior to treatment or an intervention, often including a health questionnaire and physical examination.

Histogram A graphical representation of the frequency of observations within particular categories.

Human participants Individuals who agree to take part in a study after being fully informed of the nature of the study and the risks and benefits involved.

Hyperbaria This term is used to indicate increased barometric pressure.

Hypobaria This term relates to decreased barometric pressure.

Hypotheses Suggested explanations for a group of facts which are normally then tested through investigation. Often a null hypothesis and an alternative hypothesis are presented.

Ideal gas A gas that behaves according to the ideal gas laws which specify the behaviour of gas under known conditions of volume, temperature and pressure.

Impulse A product of force and time.

Inferential data analysis An examination of data to determine with a degree of confidence relationships and differences in the populations under consideration. Relationship tests include Spearman's rank and Pearson's product moment correlation tests. Difference tests include t-test, analysis of variance, Mann–Whitney, Wilcoxon, Kruskal–Wallis, and Friedman.

Informed consent A signed agreement to take part in an activity following a full explanation of the nature of the activity and the inherent risks and benefits.

Internal energy The result of the movement of atoms within matter, often expressed as heat.

Internet A network of computers which allows the communication of electronic information.

Interval In relation to data, that which is continuous and may be subjected to parametric statistical analysis.

Introduction The section of a scientific report which provides context.

Involuntary An action which requires no conscious cognitive process.

Isokinetic A change of joint angle as a result of muscle shortening or lengthening that takes place at a constant fixed velocity.

Isometric A muscle contraction that does not result in muscle lengthening or shortening or a change in joint angle.

Isotonic A change in joint angle as a result of muscle lengthening or shortening that takes place at a variable velocity.

Joule A unit of energy or work defined as that required to displace a force of 1 N through 1 m in the line of action of the force.

Journals Publications of peer-reviewed academic work, often including original research.

Kilocalorie   A unit of energy defined as that required to increase the temperature of 1 kg of water by 1°C under standard conditions.

Kinetic energy   The capability to do work as a result of movement.

Laminar   In relation to the flow of fluids, the situation with uniform velocity throughout a vessel.

Lever   The application of a force pivoted about a fulcrum, which results in a mechanical advantage.

Line graph   A graphical representation of data using an x–y coordinate system, with successive coordinate points joined by a line.

Mean absolute deviation   A measure of variation of a sample of data, the calculation of which includes all observations within the sample of data.

Mean   A measure of central tendency of a sample of data, the calculation of which includes all observations within the sample of data.

Mechanical stress   The application of force, expressed in terms relative to the cross-sectional area of the object.

Median   A measure of central tendency of a sample of data, the calculation of which includes all observations within the sample of data, but does not account for the magnitude of each observation.

Menu-driven software   Commands are not required in order to operate this type of software. Rather, the software is operated by choosing options from a menu on the computer screen.

Methods   The section of a scientific report that describes the participants, protocol and measurements included in a study. It should provide enough information for the study to be repeated.

Mode   A measure of central tendency of a sample of data based on the most frequent observation.

Modem   A device that enables a computer to communicate with other computers through a telephone line.

Modulus of elasticity   A constant value which is specific to a material, and quantifies the elastic properties of that material.

Molarity   The number of moles of solute per litre of solution.

Mole   The amount of substance, defined as 6.02×10^{23} atoms, ions or molecules of a substance.

Mouse   A device attached to a computer which allows the movement of a cursor on a computer screen.

Multimedia   The presentation of information from a number of sources in an integrated way, including the use of video, computers and slides.

Network   A number of connected computers which can communicate with each other.

Newton   The unit of measurement of force, abbreviated as N.

Nominal   In relation to data, that which is discontinuous and consists of categories.

Normal distribution   A bell-shaped frequency distribution that appears to underlie many human variables.

Ordinal   In relation to data, that which is discontinuous and may be subjected to non-parametric analysis.

Paradigm   In relation to scientific work, a model of investigation.

Partial pressure   In relation to gases, the pressure of a single gas within a mixed gas.

Pascal   The standard international unit of pressure, abbreviated as Pa.

Performance   The ability to undertake a task.

Pie chart   A graphical representation of data in the form of a circle, where the size of each sector reflects the magnitude of the category represented.

Plagiarism   The use of ideas or work of another person, without giving appropriate credit.

Plastic   A material which deforms under loading and does not return back to its original length.

Population   In relation to statistics, the entire aggregate from which individuals or samples are drawn.

Potential energy   The capability of an object to do work due to its position.

Power   The unit of measurement for work done per unit time, or energy used per unit time, given the unit watt (W).

Pressure gradient   The difference in pressure between two points which results in the movement of a fluid (i.e. liquid or gas).

Principle of conservation of energy   The first law of thermodynamics states that energy cannot be created or destroyed, but can be changed from one form to another.

Pressure   The unit of measurement for the force exerted per unit cross-sectional area, given the unit pascal (Pa).

Protein   A nutrient composed of the chemical compounds nitrogen, carbon, hydrogen and oxygen, and which forms an essential part of the diet.

Pulmonary   A term used to describe something related to the lungs, e.g. pulmonary circulation.

Radiation   A form of heat exchange which involves the transfer of electromagnetic waves.

Random errors   Errors that are not systematic and have no consistency during the measurement process.

Range   A measure of variation of data in a sample which simply takes into account the highest and lowest value of the sample.

Ratio   The highest level of data, which is continuous, and may be subjected to parametric analysis.

References   The section of a scientific report which includes the sources of the literature consulted in preparation of the report.

Repeated measures   Data which are collected from the same participants on more than one occasion.

Research question   In a scientific study or report, the issue to be addressed.

Results   The section of a scientific report which presents the findings, either in prose or in graphical form.

Review of literature   The section of a scientific report which makes reference to previous work published in the area.

Rigid body   A body which does not deform under conditions of moderate force application.

Sample   A group taken from a population which is representative of the population under investigation.

Scalar   A quantity which has a magnitude but no direction.

Scatter plot   A graphical representation of data by plotting x–y coordinates in two dimensions.

Software   A system installed on a computer which enables particular tasks to be undertaken.

Solute   Substance that dissolves in a solvent, forming a solution.

Solution   A mixture of two or more substances, formed from a solute and a solvent.

Solvent   A liquid in which a solute dissolves.

Standard deviation   A measure of variation within a sample about a mean average value, which is numerically equal to the square root of the variance.

Standard error of the mean   A measure of variation of sample means about the mean average of the sample means.

Static electricity   A result of contact between two objects creating a net movement of positive and negative charges within each object.

Statistical error   An error in which an inappropriate conclusion is drawn from an investigation due to the criteria set for the data analysis. A type I statistical error occurs when no real difference exists but the data analysis shows a difference. A type II statistical error occurs when a real difference exists but the data analysis shows no difference.

STPD   This abbreviation refers to gas volumes under specific standardised conditions, which include standard **t**emperature (in kelvins), ambient **p**ressure (in mm of mercury) and **d**ry (no water vapour).

Strain   The extent of deformation of a material in relation to its original length, expressed as a percentage.

Strain gauge   A device for determining the extent of deformation of a material as a result of stress placed upon it.

Surface tension   In water, the result of hydrogen bonding with oxygen.

Système International (SI) units   A standardised system of units for reporting the magnitude of measurements, which is accepted internationally.

Tensile limit   The maximum load that may be applied to a material before it fractures.

Tension   The application of a force which acts to lengthen an object.

Theory   A explanation for observed events.

Thermal conductivity   The method of transfer of internal energy between atoms within a material.

Thermodynamics   The properties of a material which relate to the transfer of heat energy.

Third person   A style of writing which avoids the use of terms such as 'I', 'we', etc., and is used in scientific report writing.

Torque   The product of force and displacement of the application of the force from the pivotal fulcrum.

Trigonometry   A mathematical process which is based on the properties of triangles.

Turbulent   The erratic pattern of flow of a fluid when the velocity of flow of the fluid is high.

Variance   A measure of variation within a sample of data, calculation of which is based on all the observations in the sample.

Vector   A quantity which has both magnitude and direction.

Viscosity   An internal friction of a fluid which provides resistance to motion.

Voluntary   An action involving a conscious cognitive process.

Watt   The unit of measurement of power, abbreviated as W.

Web browser   Software that enables navigation through the contents of the World Wide Web in an orderly way.

Work   The transfer of energy, defined as the displacement of the application of the point of a force through 1 m in line with the direction of application of the force.

World Wide Web   Information on computers which may be accessed by other computers.

Yield point   The point in the deformation of a material at which the property of a material changes from elastic to plastic.

APPENDIX 1

HEALTH QUESTIONNAIRE

The purpose of this questionnaire is to find out about your health prior to participation in an exercise test. It is important that you answer the questions truthfully, and as completely as possible. All information given will be treated in confidence.

Name _____

Age _____

1 How active are you?

Do you perform vigorous activity (i.e. makes you hot and out of breath)

Less than once a month
Once a month
Once a week
Two or three times a week
Four or five times a week
More than five times a week?

2 Have you ever suffered from any form of heart complaint?

	Yes	No

3 Have you ever suffered from

	Yes	No
Asthma	Yes	No
Diabetes	Yes	No
Bronchitis	Yes	No
Epilepsy	Yes	No
High blood pressure	Yes	No
Any other condition I need to know about?	Yes	No
If yes, give details		

4 Have you had to consult your doctor in the past
three months?   Yes   No
If yes, give brief details   _____
                               _____

5   Are you currently taking any form of medication?   Yes   No
If yes, give brief details   _____
_____

6   Do you have a muscle or joint injury?   Yes   No
If yes, give brief details   _____
_____

7   Have you suffered from a bacterial or viral infection
in the last two weeks?   Yes   No
If yes, give brief details   _____
_____

8   Have you had cause not to train in the last two weeks for a
reason related to your health?   Yes   No
If yes, give brief details   _____
_____

9   Is there any reason why you should not be able to complete
tests which require maximum effort?   Yes   No
If yes, give brief details   _____
_____

Name   _____ Signed _____ Date _____

Name of Guardian* _____ Signed _____ Date _____

Exercise Physiologist _____ Signed _____ Date _____

(Note: * indicates to be completed only if the subject is under 18 years of age.)

Appendix 2

EXAMPLES OF CONSENT FORMS

**Cheltenham and Gloucester College of Higher Education
Exercise Physiology Laboratory informed consent form**

I have had full details of the tests I am about to complete explained to me. I understand the risks and benefits involved, and that I am free to withdraw from the tests at any point. I confirm that I have completed a health questionnaire, and I am in a fit condition to undertake the required exercise.

Name _____ Signed _____ Date _____

Name of Guardian* _____ Signed _____ Date _____

Exercise Physiologist _____ Signed _____ Date _____

(Note: * indicates to be completed only if the subject is under 18 years of age.)

University of Brighton informed consent form for adults

The characterisation of $\dot{V}O_2$ kinetics in children and adults

I, _____, agree to participate in a research project conducted by Dr C. A. Williams and Dr Helen Carter for the purpose of examining the responses of aerobic power in children compared to adults. The experimenter has explained to my satisfaction the purpose of the experiment and the possible risks involved. I have had the principles and procedures explained to me, I have read the information sheet and I understand them fully. I am aware that I:

1   Will have my height, weight, and skinfolds (triceps and biceps, subscapular and iliac) measured.
2   Will have my oxygen measured whilst performing a running test to exhaustion (approximately ten minutes long).
3   Will visit the laboratory on two further occasions to run at a moderate intensity three times and a high intensity twice for six-minute periods. After each run a rest period of six minutes will be given.
4   Will have a small blood sample taken from my thumb on the second and third visit to the laboratory. In total, there will be four blood samples taken.

I agree to participate in the testing and I understand that I may withdraw from the study at any time, or stop any test procedure if I experience unusual discomfort. I also understand that the staff conducting the test will discontinue the test if any indications of abnormal response become apparent.

I understand prior to the tests, that the research staff will have fully explained the procedures being used and that I will have an opportunity to ask any questions that I may have. I acknowledge that I have read this form and that I understand the test procedures and inherent risks and benefits for me due to participation in this study.

Name _____   Signed _____   Date _____

Investigator

Witnessed _____   Signed _____   Date _____

University of Brighton informed consent form for young children

The characterisation of $\dot{V}O_2$ kinetics in children and adults

I, _____ (please print parent's name), agree to my child participating in a research project conducted by Dr C. A. Williams and Dr Helen Carter for the purpose of examining the responses of aerobic power in children compared to adults. The experimenter has explained to my satisfaction the purpose of the experiment and the possible risks involved. I have had the principles and procedures explained to me, I have read the information sheet and I understand them fully. I am aware that my child:

1 Will have his height, weight, and skinfolds (triceps and biceps, subscapular and iliac) measured.
2 Will have his oxygen measured whilst performing a running test to exhaustion (approximately ten minutes long).
3 Will visit the laboratory on two further occasions to run at a moderate intensity three times and a high intensity twice for six-minute periods. After each run a rest period of six minutes will be given.
4 Will have a small blood sample taken from his thumb on the second and third visit to the laboratory. In total, there will be four blood samples taken.

I agree to allow participation of my child in the testing and I understand that I may withdraw him from the study at any time, or stop any test procedure if my child experiences unusual discomfort. I also understand that the staff conducting the test will discontinue the test if any indications of abnormal response become apparent.

I understand prior to the tests, that the research staff will have fully explained the procedures being used and that I will have an opportunity to ask any questions that I may have. I acknowledge that I have read this form and that I understand the test procedures and inherent risks and benefits for my child due to participation in this study.

Name _____ Signed _____ Date _____
(parent's/guardian's)

Investigator

Witnessed _____ Signed _____ Date _____

University of Brighton informed consent form signed by younger children

Name _____

I am happy to take part in your research project which is looking at exercise and fitness. I have had the project explained to me by Craig Williams, who has visited our school.

I understand that if I am unhappy about anything that I am being asked to do, I can say so and if I want to I can stop taking part in the project at any time.

If I am not sure of something then I can ask and Craig or Helen will explain it to me.

Signed _____   Date _____

Investigator _____

Appendix 3

THERMAL EQUIVALENTS OF OXYGEN

Thermal equivalents of oxygen for the non-protein respiratory quotient (RQ), including per cent kilocalories and grams derived from carbohydrate (CHO) and fat

Non-protein RQ	kcal per litre O_2	Per cent kcal derived from		Grams per litre O_2 derived from	
		CHO	Fat	CHO	Fat
0.707	4.686	0.0	100.0	0.000	0.495
0.71	4.690	1.0	99.0	0.012	0.490
0.72	4.702	4.4	95.6	0.054	0.473
0.73	4.714	7.8	92.2	0.095	0.456
0.74	4.727	11.3	88.7	0.136	0.439
0.75	4.739	14.7	85.3	0.177	0.423
0.76	4.751	18.1	81.9	0.218	0.406
0.77	4.764	21.5	78.5	0.259	0.389
0.78	4.776	24.9	75.1	0.301	0.372
0.79	4.788	28.3	71.7	0.342	0.355
0.80	4.801	31.7	68.3	0.383	0.338
0.81	4.813	35.2	64.8	0.424	0.321
0.82	4.825	38.6	61.4	0.465	0.304
0.83	4.838	42.0	58.0	0.507	0.287
0.84	4.850	45.4	54.6	0.548	0.270
0.85	4.862	48.8	51.2	0.589	0.254
0.86	4.875	52.2	47.8	0.630	0.237
0.87	4.887	55.6	44.4	0.671	0.220
0.88	4.899	59.0	41.0	0.712	0.203
0.89	4.911	62.5	37.5	0.754	0.186
0.90	4.924	65.9	34.1	0.795	0.169
0.91	4.936	69.3	30.7	0.836	0.152
0.92	4.948	72.7	27.3	0.877	0.135
0.93	4.961	76.1	23.9	0.918	0.118
0.94	4.973	79.5	20.5	0.959	0.101
0.95	4.985	82.9	17.1	1.001	0.085
0.96	4.998	86.3	13.7	1.042	0.068
0.97	5.010	89.8	10.2	1.083	0.051
0.98	5.022	93.2	6.8	1.124	0.034
0.99	5.035	96.6	3.4	1.165	0.017
1.00	5.047	100.0	0.0	1.207	0.000

Source: G. Lusk (1926), *Science of Nutrition*

Appendix 4

SCIENTIFIC JOURNALS IN SPORT AND EXERCISE

American Journal of Sports Medicine
Applied Psychophysiology and Biofeedback
Australian Journal of Science and Medicine in Sport
British Journal of Physical Education
British Journal of Sports Medicine
Canadian Journal of Applied Physiology
Culture Sport Society
European Journal of Applied Physiology
European Physical Education Review
Exercise and Sports Science Review
International Journal of Physical Education
International Journal of Sport Psychology
International Journal of Sport Nutrition
International Journal of Sports Medicine
International Journal of the History of Sport
International Review for the Sociology of Sport
Journal of Ageing and Physical Activity
Journal of Applied Biomechanics
Journal of Applied Physiology
Journal of Applied Sport Psychology
Journal of Human Movement Studies
Journal of Motor Behaviour
Journal of Physical Education Recreation and Dance
Journal of Physiology
Journal of Science and Medicine in Sport
Journal of Sport and Exercise Psychology
Journal of Sport and Social Issues
Journal of Sport Behaviour
Journal of Sport Rehabilitation
Journal of Sports Management
Journal of Sports Medicine and Physical Fitness
Journal of Sports Sciences
Journal of the Philosophy of Sport
Medicine and Science in Sport and Exercise
Motor Control

New Studies in Athletics
Pediatric Exercise Science
Perceptual and Motor Skills
Quest
Research Quarterly for Exercise and Sport
Sociology of Sport Journal
Sport History Review
Sport Psychologist
Sports Medicine
Women in Sport and Physical Activity Journal

Useful addresses

American College of Sports Medicine
P.O. Box 1440
Indianapolis
IN 46206-1440
USA

British Association of Sport and Exercise Sciences
114 Cardigan Road
Headingley
Leeds LS6 3BJ

British Olympic Medical Centre
Northwick Park Hospital
Watford Road
Harrow
Middlesex HA1 3UJ

Lilleshall Sports Injury and Human Performance Centre
Lilleshall National Sports Centre
Near Newport
Shropshire TF10 9AT

The National Sports Medicine Institute of the United Kingdom
c/o Medical College of St Bartholomew's Hospital
Charterhouse Square
London EC1M 6BQ

Useful web sites

http://www.acsm.org/sportsmed/   The web site of the American College of Sports Medicine.

http://www.ausport.gov.au/aismenu.html   The web site of the Australian Institute of Sport based in Canberra.

http://www.cid.ch/DAVID/mainmenu.html   A very good web site for anatomy with many images from magnetic resonance imaging scans.

http://www.gla.ac.uk/Library/Depts/MOPS/Stats/medstats.html   A very comprehensive web site for medical statistics.

http://www.gssiweb.com/   The web site of the Gatorade Company with many links to nutrition and fluids in sport and exercise.

http://www.hea.org.uk/   The web site of the Health Education Authority.

http://www.mspweb.com/orgs.html   An excellent web site which acts as a link to many other sport and exercise, medicine and rehabilitation web sites.

http://www.newscientist.com/   The web site of the popular science periodical *New Scientist*.

http://www.nsmi.org.uk/   The web site of the National Sports Medicine Institute.

http://www.physsportsmed.com/   The web site of the journal *The Physician and Sports Medicine*.

http://www.sportsci.org/   An informative web site dedicated to sports science and statistics.

Appendix 5

MEASUREMENT CONCEPTS

In exercise and sport work, either in the laboratory or in the field, it is often necessary to critically examine any data collected. Quite often it is desirable to know how similar the results would be were the same procedure repeated on more than one occasion. Such a concept may be given several terms depending on the type of measurement. If measurements are taken from a single piece of equipment, the appropriate term is **precision**. If measurements are taken from more than one piece of equipment, and then used to determine a further value, the appropriate term is **reliability**. If data are taken from a psychological inventory, the appropriate term is also reliability.

It is also desirable to know whether measurements are giving the correct information. Again, such a concept may be given several terms depending on the type of measurement. When data are collected from a single measurement, the appropriate term is **accuracy**. When data are derived from several measurements, the appropriate term is **validity**. When data are derived from a psychological inventory, the appropriate term is also validity.

When dealing with data from interviews, it is not possible to separate reliability from validity. A composite measure is that of **trustworthiness**. Saying data are trustworthy is similar to saying they are both valid and reliable.

An example of precision without accuracy

An example of accuracy and precision

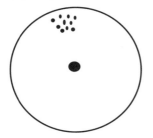

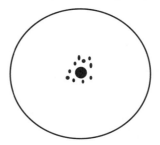

	Reliability	*Validity*
Single measurement	Precision	Accuracy
Compilation of measurements	Reliability	Validity
Psychological inventory	Reliability	Validity
Interview	Trustworthiness	

INDEX